DANGER, DEATH AND DISASTER
IN THE CROWSNEST PASS MINES 1902–1928

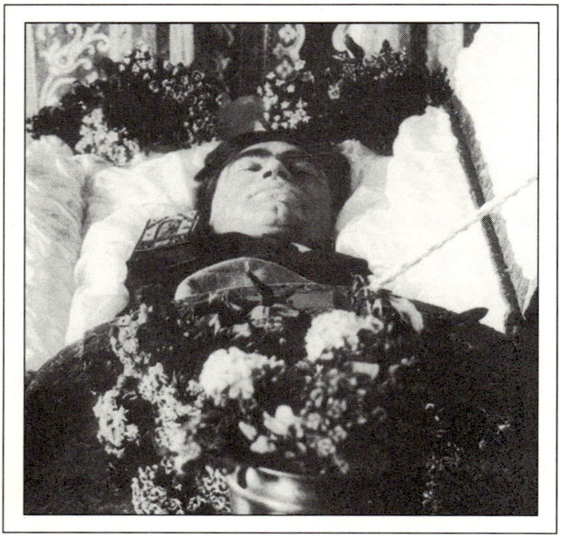

DANGER, DEATH AND DISASTER
IN THE CROWSNEST PASS MINES 1902-1928

KAREN BUCKLEY

UNIVERSITY OF CALGARY PRESS

Published by the
University of Calgary Press
2500 University Drive NW
Calgary, Alberta, Canada T2N 1N4
www.uofcpress.com

No part of this publication may be reproduced, stored in a retrieval system or transmitted, in any form or by any means, without the prior written consent of the publisher or a licence from The Canadian Copyright Licensing Agency (Access Copyright). For an Access Copyright licence, visit www.accesscopyright.ca or call toll free to 1-800-893-5777.

We acknowledge the financial support of the Government of Canada through the Book Publishing Industry Development Program (BPIDP), the Alberta Foundation for the Arts and the Alberta Lottery Fund—Community Initiatives Program for our publishing activities.

ALBERTA LOTTERY FUND

Canada

Canada Council for the Arts   Conseil des Arts du Canada

∞ This book is printed on acid-free paper
Cover design, page design and typesetting by Mieka West. Printed and bound in Canada by AGMV Marquis

© 2004 Karen Buckley

Cover photo courtesy Glenbow Archives, NC-54-836.
Inside cover photo by Karen Buckley.

Library and Archives Canada Cataloguing in Publication

Buckley, Karen Lynne, 1963-
 Danger, death and disaster in the Crowsnest Pass mines, 1902-1928 / Karen Buckley.

Includes bibliographical references and index.
ISBN 1-55238-132-3

1. Coal mines and mining—Crowsnest Pass Region (Alta. and B.C.)—History—20th century. 2. Coal mine accidents—Crowsnest Pass Region (Alta. and B.C.)—History—20th century. 3. Crowsnest Pass Region (Alta. and B.C.)—Social life and customs—20th century. 4. Crowsnest Pass Region (Alta. and B.C.)—Social conditions—20th century. 5. Crowsnest Pass Region (Alta. and B.C.)—History—20th century. I. Title.

TN806.C22C76 2004   971.23'4   C2004-904311-0

As always, for my parents.

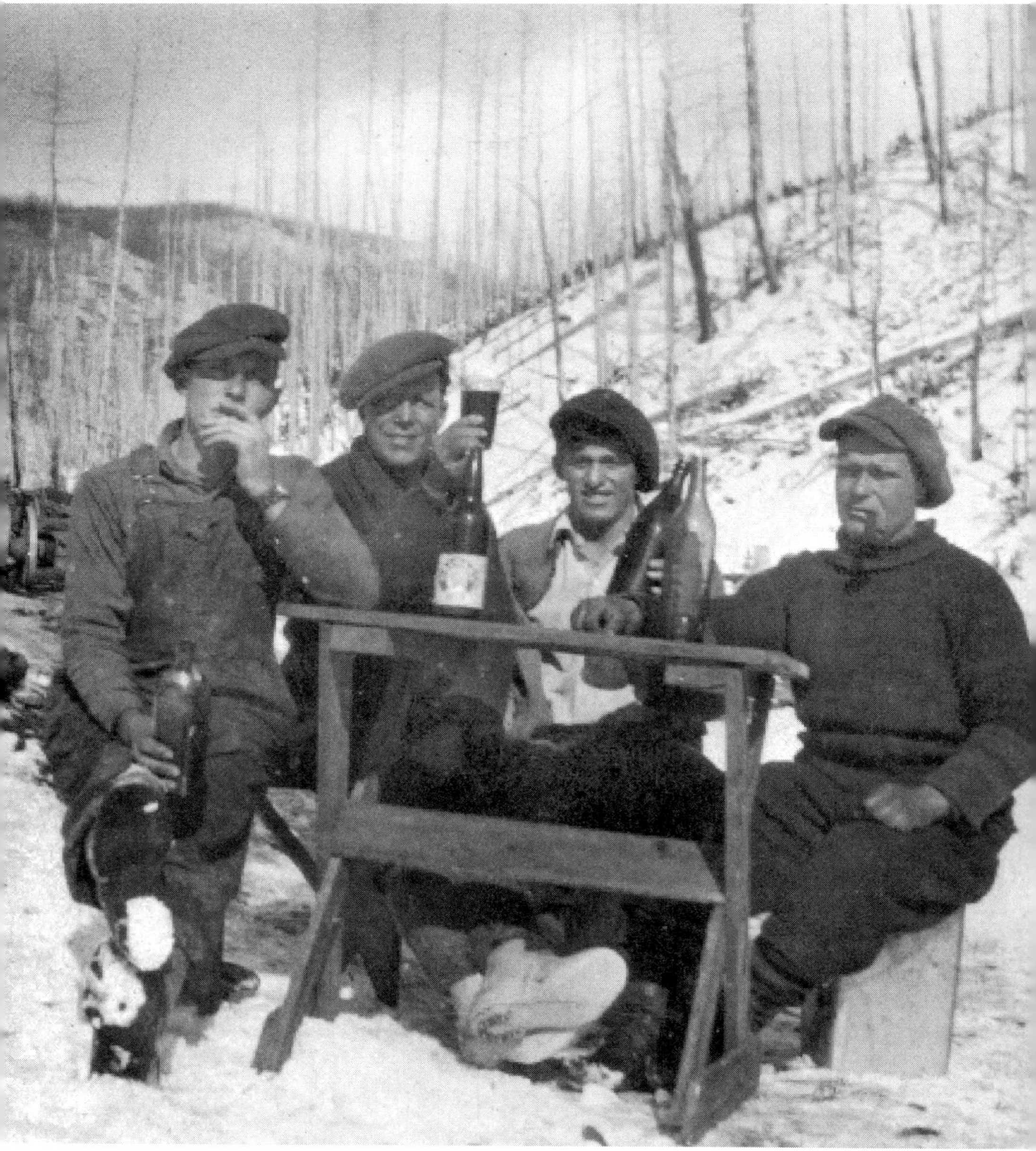

CONTENTS

List of Photos viii
Acknowledgments ix
Introduction xi
1: Danger 1
2: Death 33
3: Disaster 75
Conclusion 141
Appendix A: Tables and Charts 147
Appendix B: Mining Definitions 153
Bibliography 155
Notes 165
Index 187

LIST OF PHOTOS

1. Hillcrest Mine Disaster, Families waiting on Hillside, 1914
2. Draegermen with Equipment Rushed to Scene of Bellevue, 1910
3. Convention of Alberta Undertakers in Edmonton, 1907
4. Hearse and Driver (probably Joe Plante, T.W. Davies' assistant), Connor Tiberg's Funeral, c. 1916
5. Women with Casket, c. 1928
6. Man in Open Casket, c. 1928
7. Funeral gathering at Train Station, c. 1916
8. Miners posing with Davey Safety Lamps, 1910
9. Mourners with Coffins after Mine Disaster, 1914
10. Funeral of Frank Malec, 1923
11. Funeral Procession, Main Street Blairmore, c. 1924–30
12. Mass Graves at Hillcrest Cemetery
13. Mass Graves at Fernie Cemetery
14. Grave Marker of John Hovan, Fernie Cemetery
15. Grave Marker of James Mitchell, Fernie Cemetery
16. Grave Marker of Thomas Bardsley, Hillcrest Cemetery
17. Grave Marker of Henry Plasman, Coleman Union Cemetery
18. Grave Marker of Joseph Homenick, Coleman Roman Catholic Cemetery
19. Grave Marker of Henry and Harry Grewcutt, Coleman Union Cemetery
20. Grave Marker of Samuel Edwin, Hillcrest Cemetery
21. Grave Marker of George Lothian, Coleman Union Cemetery
22. Grave Marker of Robert Smith, Hillcrest Cemetery
23. Grave Marker of Charles Coats, Hillcrest Cemetery
24. Grave Marker of Jules Falip, Fernie Cemetery
25. Grave Marker of William Clarke, Fernie Cemetery
26. Grave Marker of Fred Alderson, Hosmer Cemetery
27. Hillcrest Mine Disaster Memorial, Hillcrest Cemetery
28. Hillcrest Mine Disaster Memorial, Hillcrest Cemetery

ACKNOWLEDGMENTS

I owe a debt of gratitude toward many people for their assistance during the writing of this book. First to Dr. Herman Ganzevoort for his encouragement, enthusiasm, and his sense of humour; to the knowledgeable staff of the Glenbow Archives, the Provincial Archives of Alberta and the Provincial Archives of British Columbia; to Mike Pennock of the Fernie and District Historical Society, Sr. Leona Henke of the Roman Catholic Diocese Archives, and Apollonia Steele of the University of Calgary Special Collections; to the University Research Grants Committee and to Dr. Frits Pannekoek for assistance with research time and funding; to the editorial staff at the University Press for their patience and professionalism; to the people of the Crowsnest Pass, especially to the dedicated staff of the Frank Slide Interpretative Centre and to Roy Lazzarotto for the informative and personal tour of the Bellevue mine and subsequent interview. Many thanks are also due to Diane Peterson for her personal interest in the subject, her knowledge of the area, and above all just for herself. And for my parents who have supported me in so many ways for so long; there are no words adequate to say thanks.

INTRODUCTION

On the eastern edge of the Coleman Union Cemetery stands a tall grey marble obelisk. Written on its side is the inscription:

> In
> Loving Memory of
> Henry Grewcutt
> Who was
> Accidentally
> Killed in a Mine
> Explosion at
> Coleman, Alta
> April 2, 1907
> Aged 47 Years
> Also His Son
> Harry
> Killed at the Same Time
> Aged 21 Years
> 3 Months.

On 2 April 1907, Henry Grewcutt and his son Harry had reported for their regular shift in the McGillivray Mine at Coleman. The Grewcutts collected their safety lamps from the lamphouse and walked with the other miners to the coalface. As father and son entered the room where they worked together, a terrific explosion shook the mine. In the ensuing moments of panic, the miners followed their instinct and made a dash for open air, trying to outrun the flow of deadly afterdamp or methane gas. In the rush to the surface, Harry became separated from his father, made a wrong turn and fell down a coal chute. Henry managed to reach the mine entrance, but when he found that his son was not among the crowd of gasping, coughing miners, he went back into the mine. Rescue workers eventually pulled eleven men from the mine that morning. Eight of the unconscious miners were lucky and revived from the effects of gas in the open air; Henry and his son Harry both died. The third man killed that day was Charlie Hutton, the fireboss on the shift, who had rushed into the mine at the first sound of the explosion in a desperate effort to help the men inside.[1]

In his report on the accident, the Chief Inspector of Mines "attributed the explosion to the carelessness of Hutton and of Grewcutt and in both of them not examining the working place as they ought to have done."[2] The inspector reasoned that the gas causing the explosion had been present before the shift and that Hutton, as fireboss, should have found it on his own rounds. Also, although both the Grewcutts carried safety lamps, Henry relied on an open light as he entered his section of the coalface. Many miners preferred the greater illumination that an open light gave them. However, in this instance, the naked flame ignited the pocket of gas that had gone undetected near the coalface.

After the coroner had examined the dead, a hastily-put-together union committee took charge of washing and preparing the bodies for burial. United Mine Workers of America members and other miners later marched in the large funeral procession to the gravesite where the newly appointed president of Local 2633 read the service, a duty Henry Grewcutt had performed in the past as the mining union president. Father and son were buried together in a double plot. After the funeral, there is an indication that the grieving widow and mother thought of suing the McGillivray mine company for damages,[3] but apparently nothing came of this. Mrs. Grewcutt is also mentioned in the minutes of the Coleman Local as she sought some form of compensation for her double loss, but she then disappears from the record. She was not buried with her husband and son and may have left town after the tragedy.

The Grewcutts' lives and their deaths were similar to those of many others who came to the Crowsnest Pass[4] in the days when coal was needed by a growing nation and hands to dig the coal were needed even more. Father and son were two members of the growing population in the Pass, attracted there by the employment opportunities offered by a busy string of coal mines that ran from Bellevue and Hillcrest on the Alberta side of the border to Coal Creek and Morrissey in British Columbia. The Grewcutts had lived in town long enough to put down roots and create ties to the community. Henry Grewcutt not only held the post as the first Sunday school superintendent at St. Paul's Church, he was also the current president of the Coleman Local of the United Mine Workers union.[5] At twenty-one, Harry probably spent his time away from the mines enjoying any one of the numerous dances and social or sporting events available to residents of the area.

Harry followed the coal-mining tradition and joined his father at the coalface. Although neither Grewcutt could help but be aware of the dangers present in the mines, each chose to respond to that danger in his own way. Both died in the mine, even though other miners rushed to the sound of the explosion hoping to administer first aid or assist in a rescue attempt. A doctor was called, the bodies identified, and Mrs. Grewcutt notified of her loss. A coroner held an inquest and an undertaker assisted with the arrangements for the funeral. Father and son were mourned and buried. Mrs. Grewcutt chose to erect a large grave marker to memorialize not only the dates of their death but also the facts of their dying. She initially responded to the tragedy with grief and anger, but then disappeared into anonymity.

This story was repeated many hundreds of times in the Pass. In 1902, 128 men died in the explosion at Coal Creek, the mine located five kilometres from Fernie. Smaller explosions resulted in the deaths of four men at Morrissey in 1903, seven at Michel and an additional fourteen at Morrissey in 1904. Thirty-one died at Bellevue in 1910 followed by the massive loss of 189 men at Hillcrest in 1914. A year later there were another thirteen fatalities at Michel in 1916, thirty-four at Coal Creek in 1917, ten more at Coleman in 1926 and finally six dead at Coal Creek in 1928 (see Table 1 and Chart 5). Four hundred and fifty-one men died in these and additional lesser disasters, but it was a rarity when the local papers did not report at least one accident. As required by law, the mine owners throughout both provinces reported these accidents to the Mines Branch, using the headings of Fatal, Serious or Slight. On average, the Serious and Slight accidents accounted for at least 110 accidents a year, or two per week between 1906 and 1928 (Chart 4).

Given this loss of life and the weekly, if not daily, tally of accidents both major and minor, how did the individuals and communities of the Crowsnest Pass respond to the constant threat of danger and the realities of death and disaster? In the early part of the twentieth century, disaster relief agencies, grief counsellors, or other crisis intervention professionals did not exist. Instead, the inhabitants of the Pass depended on their own resources, their families, or the organizations and people in their community to help them survive trauma and loss. Most found a way to cope; others did not or could not.

The geographical area of the Crowsnest Pass runs from Bellevue and Hillcrest in Alberta to Morrissey in British Columbia. The division created by the provincial boundary is wholly artificial in the sense that the Crowsnest Pass is essentially a homogenous area in terms of geography and economy. Since the Pass towns shared a single common industry, the problems and concerns they faced were correspondingly similar if not identical. The universal dangers and daily routines effectively made the Pass one community where residents provided support in experiences both happy and tragic. Grace Dvorak of Coal Creek remembered: "We shared the feelings of uncertainty and loss of the family who had lost a loved one, knowing full well it could have been us in their position. That was the way of life in a mining town."[6] Ferucio DeCecco of Coleman was only one of many miners who had fond memories of the "friendship and fun down in the mine."[7] Miners marched together when called to Union action, while rescue crews and supplies were regularly rushed to other mines in times of trouble. On happier occasions, picnics, dances, sporting events, and community fundraisers served to solidify ties throughout the Pass by bringing the area inhabitants together.

One of the common elements shared throughout all the towns in the Crowsnest Pass was the diversity of ethnic groups. One early resident of Coleman remembered that "there were four separate districts each segregated from the others. They were Slav Town, Dago Town, Bush Town (mainly Belgians and Russians) and the aristocrats [in the] English district."[8] A schoolteacher stated: "Bellevue School was a League of Nations ... 30–40 countries in one room,"[9] while Fernie was such a virtual Tower of Babel that in 1912 it even earned a seat in town for an Italian Consul.[10] Census figures for the Crowsnest Pass show that although the towns were usually at least 50 per cent English-speaking, the total number of Italians, French, and eastern Europeans rivalled that of their English neighbours. Immigration patterns also show that as the years went on, the number of ethnic groups began to displace the British population (see Charts 1, 2, and 3).

Some early residents remembered that there were notes of discord among the differing ethnic groups in the pass. Wilfred Eggleston of Bellevue remembered: "Of course I know there were undercurrents of discord and friction in the Pass ... ," while William Park recalled that "the West Canadian Collieries ... employed a large number of Frenchmen who were not always easy to get along with."[11] Ill will regarding ethnic differences could be covert, as when two black miners were assigned to work in the driest and dustiest portion of the Coal Creek mine,[12] or the discord could flare up in a public display, as when Constable Stevens of Fernie was run out of town by the miners. An ill-advised and ill-timed remark of his was construed to mean that Stevens wished more of the ethnic population had died in the Coal Creek disaster in 1902.[13]

However, the majority of residents of the Crowsnest Pass recalled a great deal of fun and laughter. New Ukrainian residents of Coleman, "not knowing the English language ... grocery shopped by signs and hand motions receiving remarkable results from the store keepers with laughter from both sides."[14] A hundred members of the Slavic community attended the wedding of one of their own, but "the English-speaking residents of Coleman were [also] there in full force."[15] The Locals of the United Mine Workers of America in the Crowsnest Pass were especially diligent in their attempts to bring all ethnic divisions into one homogenous group, albeit for their own purposes. Over all, with the passing of the years and the emergence of each new generation, the ethnic divisions tended to soften and blur. While the Crowsnest Pass was remarkable for its ethnicity, at one point counting forty-six different groups,[16] ethnicity did not become the sole defining character of the area.

The personal choices miners made could sometimes exacerbate the dangers in their work. Like the Grewcutts, many miners took risks by choosing to resist the use of safety lamps or of "permitted" explosives in favour of quicker but more dangerous mining methods.

Environmental and workplace hazards over which the miners had little control included the unpredictable bumps and outbursts of methane gas that could occur at any time. The Coal Creek and Morrissey mines were especially notorious for bumps and blowouts; between 1925 and 1928, forty-two significant disturbances were experienced although "hundreds had occurred over a period of 20 years."[17]

Miners developed various responses to the constant presence of these hazards, although none of these reactions were what could be termed a "blanket" response or one that all miners developed as a group. Rather, a determination of certain patterns in individuals' reactions is possible. Some miners chose to leave the area entirely or at least left the mines to work at other jobs in town. Men also used the limited avenues available to them to agitate for safer conditions in the mines, with mixed results. Several psychological defences were employed by the miners to guard against being overwhelmed by their occupation. Some of these defences included a sense of fatalism, or of choosing to celebrate pride of craft, camaraderie, and independence in their work instead of dwelling on the constant atmosphere of danger.

In the community as a whole, various groups and individuals also developed professional and specialized responses to the presence of danger, death, and disaster. The loss of lives through mining accidents and major disasters prompted governments, mine officials, and miners to gradually become more organized in mine rescue and first aid training. Legislation in British Columbia became effective in 1909 providing for the establishment of mine rescue stations and requiring mines to keep specific types and numbers of rescue equipment on hand. In Alberta there were no regulations until 1912, when a mine rescue station was established at Blairmore with the costs borne equally between the provincial government and the mine companies.[18] However, all training with the equipment and in mine rescue techniques continued to be on a voluntary basis; the effectiveness of this response can certainly be questioned given the problems that developed in the aftermath of major explosions.

Where the work of the mine rescue workers ended, the response of another specialized group in the community took over. The positions and responsibilities of undertakers became increasingly visible in the Crowsnest Pass during this period as these men gradually became true professionals. They evolved from merchants, contractors, and erstwhile politicians (Fernie undertaker George Thomson served a term as mayor) with a sideline occupation in undertaking to become instead "Funeral Directors and Embalmers" who were charter members of their profession with the training and skills to match. The undertakers played a vital part in the recovery of bodies from accidents and disasters and in assisting individuals and the community to come to terms with death.

When death occurred on a massive scale, the disaster created a ripple effect of responses. For individuals, these included the immediate fear of hearing the mine whistle blow signifying a major incident, the trauma of waiting for news, and the psychological and physical ordeals of identifying the bodies and preparing them for burial. Communities reacted as a group to disasters in a variety of ways that included the call for relief fund subscriptions, the need to create and identify "heroes," and the immediate and mostly automatic desire to contribute physical, social or spiritual assistance to whomever was in need. Fraternal organizations, mutual benefit and ethnic societies, churches and unions all responded to disasters in various ways and to greater or lesser extents. The very last element of response to death and disaster is the memorialization of the dead. Communities and individuals chose either to commemorate the loss of loved ones in private and therefore undocumented ways, or to publicly make note of the anniversaries of the tragic events.

An examination of all these elements creates a clearer understanding of how a community develops in the presence of danger in the daily lives of its members. Miners and their families all knew, on some level, that their work in the mines might cost them their lives. When death did occur, the community reacted in several stages by drawing on the resources available to them and relying on the professional response from some of the integral members of their community. These responses all provide some insight into how mining communities coped with the presence of danger, the reality of death, and the possibility of disasters.

Source Material

Written primary source materials such as diaries, letters, and memoirs are sadly lacking for the Crowsnest Pass. The reactions of individuals to family or community tragedies are deeply personal and as such are all too infrequently noted in the official record. There are several possible explanations for the difficulties experienced in finding personal recollections. The compensation claims for the Crowsnest Pass reveal that the widows and survivors processed their claims from such diverse locations as Italy, Poland, Finland, France, and the Ukraine. These individuals responded to their tragedy by leaving the area, with the result that their voices have all but vanished.

Another reason for the relative silence of survivors is that in many cases they have left their sons and daughters to write or otherwise record the histories or recall the memories. However, the children might have been too young to remember the actual events, were never told of the traumatic times in their parents' lives, or told only the basic facts that an explosion happened and men died. This may distort the memory, or become only a retelling of a family story and not the true memory of the experience itself. Despite this, the perspective gained from these sources can be equally instructive, since it suggests that for those most deeply affected by the tragedy, the disasters were only *one* moment and not the *defining* moment that shaped their lives. Or perhaps this was a defining moment, but the memory became permanently repressed or buried.

Some oral and community histories can also reveal much when the familial continuity of the area is taken into account. Many of those interviewed in the 1960s, 1970s, and 1980s came from families who had lived in the Pass since the early part of the century. Mining was the family business and the majority of the interviewees were men who had followed their fathers and grandfathers into the mines. These men had consequently absorbed the traditions, historical behaviours, and thought processes of those before them. Their recollections are therefore not solely their own, but those of previous generations as well.

Oral histories can also reveal a great deal when combined with other sources such as photographs. A photographer at the 1914 Hillcrest mine disaster captured the townspeople sitting on a hillside, waiting for the bodies of their sons, fathers, brothers, or husbands to be brought out of the mine (see Photo 1). Many of the women are dressed in white with children appearing to sit quietly at their sides. Other groups of people stand quietly looking out of the picture. Even knowing the context of this photograph, the imagination could still easily interpret this as a Sunday gathering, perhaps a picnic in the countryside. There is no palpable sense of urgency, or desperate anticipation that one would expect to see in the faces and attitudes of the people waiting for the dead to be brought out. The photograph only "comes to life" when combined with the words of a woman who was a young child at the time. There was extreme emotion in her voice as she described the scene that the photograph captures: "... the citizens [were] sitting on the hillside *just waiting* for *some* word"[19] (emphases added). Her stark recollections do more than anything else could to suggest the depth of hidden emotions present at that scene.

The similarity of mining experiences through space and time can also serve to extend the sources for research material; the knowledge and experience of danger and death in the mines was not limited only to the towns of the Crowsnest Pass. When ten men died in an explosion at Coleman in 1926, a letter sent from Drumheller illustrated the ties shared by all miners:

> Many here, who have suffered the pangs of bereavement and suspense in similar tragedies in the past, realize more fully than the average person just what such a calamity means in a mining community.[20]

Nor was this commonality limited by provincial or even national boundaries. Oral histories gathered from early Colorado coal miners indicated that the "miners had labored under a specific set of working and social conditions and this created the 'coal community.'"[21] From the similar reminiscences and common events in these miners' lives, "it became clear that they were part of a community of Western coal miners defined by occupation, common experiences, and shared attitudes, values and expectations."[22] Nor was it only attitudes and values that were shared in this larger mining community. Mine rescue operations from Coal Creek to Nova Scotia to New Mexico were virtually identical, as were the stories of each community's response to mining disasters.

1   Hillcrest Mine Disaster, Families waiting on Hillside, 1914
    Provincial Archives of Alberta, 86.425

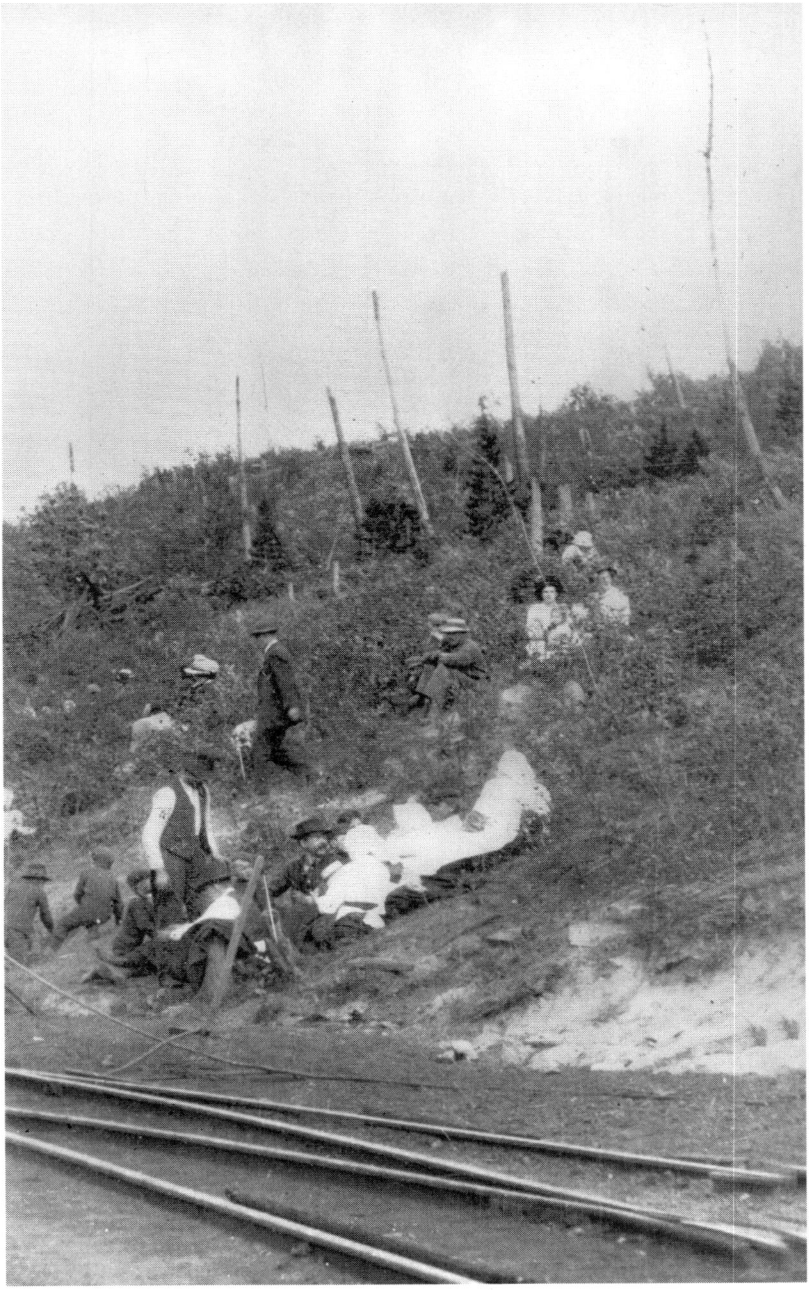

The commonality of these responses is also not limited to one period in time. Despite changes in mining methods, labour agreements, and market forces, many essential elements of underground coal mining have remained the same. When examining the formal investigations into the Coal Creek, British Columbia, explosion of 1902, and the Westray Mine, Nova Scotia, explosion ninety years later, it is surprising to note that not much has changed at a basic level. The attitudes of the miners toward their jobs, warnings of danger ignored in the push for production, descriptions of rescue attempts, the aftermath of grief; each of these are common elements that can be easily identified in mining disasters separated by a century.

One mining community is described as:

> Minetown, then, was a small isolated community of single industry characterized by close face-to-face social relations and bonds of mutual assistance. The entire work force and population depended directly or indirectly upon coal mining…. Minetown may be viewed as a community highly involved with problems of danger, injury, and death.[23]

Although the description here is from Springhill, Nova Scotia, in 1958, these comments could just as accurately have described Coal Creek in 1902 or Hillcrest in 1914. However, when using sources outside the historical timeline of the Crowsnest Pass, caution must be exercised. Despite the similarities evident between the investigations into Coal Creek and Westray, the ninety years of experience and time that separate them cannot and should not be ignored. When sources such as these were consulted, they were used with care and only when circumstances appeared to warrant drawing a direct parallel.

Mining songs can also be an important source of insight into the reactions of individuals and communities to death and disaster.[24] As song collector John O'Donnell asserts, there is an abiding universality to mining songs:

> The fact that a song is credited with an eastern Canadian source, however, does not mean that it is not known in other Canadian coal mining areas ... through much of Canada's history, coal miners were forced to seek work wherever it could be found.[25]

The mobility of the Crowsnest Pass workforce lends credibility to O'Donnell's observation. Miners came to the Pass not only from Nova Scotia but also from the United States, Britain, and Eastern Europe. Many had worked in several other mines, rubbing shoulders and becoming acquainted with each other's histories before eventually settling in the Pass. Mine Inspector Evan Evans was one of several hundred other Welshmen who emigrated to Fernie; Edgar Ash, who joined many others from Nova Scotia when he moved first to Frank and then Coleman, "knew many songs and folk tales of [that] area and was fond of singing and relating [them]."[26]

Songs can be an expression of emotion at times of great stress, as exemplified by the disaster narratives memorializing the Avondale, Lick Branch, Fraterville, and Cherry Mine disasters.[27] Mining songs also give a glimpse into the psychology of the mining experience and the response by the miners to dangers daily faced. "Don't Go Down in the Mine, Dad," is a plaintive plea by a child afraid for his father that could have been heard in any one of North America's mining towns. "Only a Miner" has been described as "an excellent demonstration of song as a reflector of a community's varied response to industrial death."[28] Songs are initially given voice by an individual's response to a tragedy, but this voice can then be amplified and shared by an entire community in a cathartic reaction.

Source material on the psychology of miners and their reaction to daily dangers is relatively scarce. A pioneering study into the psychology of human response to disasters in the 1940s studied survivor reactions after the Cocoanut Grove nightclub fire in Boston that killed 492 people.[29] This study is one of the earliest to note civilian post-traumatic stress disorder. Unfortunately, other studies specific to miners surviving entombment after coal falls or explosions are significantly scarcer.[30] In contrast, general disaster studies are abundant. There is an inescapable human fascination with true stories of mass death and destruction. Mining related disasters such as Aberfan in Wales and Buffalo Creek in the Appalachians have produced numerous works,[31] while disaster studies themselves came into their own as a science in the post-World War II years.[32]

All general disaster studies must be treated with some caution when applied specifically to mining. While a disaster such as the Frank Slide of 1903 qualifies as an astounding visual event that overwhelmed the senses of everyone who saw and experienced it, in contrast, the vast majority of mining accidents and major explosions occur underground. The extreme destruction that explosions could wreak inside a mine would only be evident on the surface by some broken buildings or a pall of smoke issuing from the entrance to the mine. Also, the psychological reactions of individuals over the uncertain whereabouts of loved ones caused by floods, mudslides, hurricanes, or tornadoes are certainly different from those experienced during a mining disaster. A trapped miner could still take comfort from the knowledge that his family was alive and well, while the family waiting anxiously for his return at least knew that the man was in the mine, even if they were uncertain whether he was living or dead.

Given the daily dangers of mining life, a differentiation must also be made between the completely unexpected, sudden disastrous situations and what might at some level be an expected event. "Death in the mines offers at least a thin layer of insulation, if only in the sense that one's nerves are braced in anticipation and one's imagination has had a chance to rehearse the possibilities."[33] However, although the loss of one or two men might arguably be anticipated, the loss of 189 such as occurred at Hillcrest in 1914 could hardly be foreseen. This disaster certainly qualifies as an event that gives "a blow to the psyche that breaks through one's defenses so suddenly and with such brutal force that one cannot react to it effectively."[34]

The pervasive public image of the coal miner as a romantic figure in source material should also be taken into account at all times. The media have developed this image for over a century to such an extent that it has nearly become what could be regarded as a "socially reinforcing myth."[35] The *Calgary Daily Herald* celebrated the stoic bravery of the Hillcrest mine rescue workers in 1914 as "square-jawed muscular grimy men who go about their tasks as if it were but part of the day's work."[36] Miners are also often described or thought of as the "archetypal industrial laborer", one who has "symbolized the ideals of solidarity, toughness, determination, and militancy."[37] Mine organizations sometimes reinforced this image, as evidenced by a widely circulated story told in 1906 by John Mitchell, president of the United Mine Workers of America. A Pennsylvanian miner was rescued after being trapped for five days after a fall of coal.

According to Mitchell the miner was unfazed by his experience and his first question to his rescuers concerned what day it was. The only worry that the trapped miner had experienced was that he might be missing the chrysanthemum show.[38]

This story also serves to illustrate an important point when trying to analyze source material for the responses of miners to danger and death. A common perception in writing about miners is to define them as the "'other,' as objects waiting to be defined rather than subjects capable of defining themselves."[39] However, miners were not two-dimensional characters, coated in black, toiling in the darkness, rising to the surface to sleep, only to descend again into the bowels of the earth. Although their occupation happened to be mining, these men had wives and families and sweethearts; they had myriad interests, skills, and hobbies; they debated politics and sports as passionately in the bars as they did when hashing over the minutiae of their day. William Anthony from Fernie spent his free time involved in community organizations; not only was he a member of the local of the United Mine Workers, he was also a member of the Knights of Pythias and the Oddfellows Lodge.[40] Jack Maddison spent the majority of his time away from the Hillcrest mines fishing and duck hunting like many of his fellow miners.[41]

Pictures and photographic histories of mining have only served to strengthen the image of the miner as one set apart from the rest of society. A pictorial article on the imminent closure of a present day mine in Cape Breton is instructively titled "One Last Whistle" and includes a full-page photograph of two miners exuding the qualities of endurance and strength.[42] Lawrence Chrismas has published two pictorial works representing miners as rugged, salt-of-the-earth types. The visual images are reinforced by Chrismas's choice of where and how he photographed the miners' portraits, and by a few well-chosen comments on their lives and their trade.[43]

This external perception of miners as a "collective" or as "the other" can sometimes serve to distort the personal recollections of individual miners; their sense of themselves and their role in the industrial world is not necessarily that held by others. To reach behind the portraits and the public image is difficult given the scarcity of personal sources and interviews. When fashioning an argument from sources both limited and brief, the following caution must always be kept in mind: "As conversations and bits of conversations became data, they were freed from the context of time, mood, dialogue, and even person."[44] Any arguments or conclusions taken from memories of miners must be reconstructed with care and attention to the true meaning and context behind the words.

1 DANGER

In the whole of the British Empire there is no occupation in which a man may meet his end in so many diverse ways as this one. The coal mine is the scene of a multitude of the most terrifying calamities.... The hydrocarbon gas which develops so freely in these mines, forms, when combined with atmospheric air, an explosive which takes fire upon coming into contact with a flame and kills everyone within its reach. Such explosions take place, in one mine or another, nearly every day.[1]

Frederick Engels, commenting on coal mining in 1844, understood mine explosions in a remarkably precise way for his time; his lament that they happened "nearly every day" accurately described the dangers faced by coal miners everywhere. Mining was a hazardous occupation in all parts of the early twentieth-century industrial world. Between 1900 to 1910, an average of one major explosion per year killed one hundred or more miners in the United States. A total of 3,241 deaths made 1907 the worst year in American coal mining history, with the Monongah, West Virginia mine explosion alone killing 361 men.[2] Similar statistics are available from Europe. While in Britain fatalities peaked at 1,484 in 1866, one 1906 disaster alone in France took 1,099 lives.[3] Although the fatality rate in the Crowsnest Pass never reached such epic proportions, the unpredictable and dangerous mining conditions took their toll on miners' lives.

In 1898 William Fernie and Lieutenant-Colonel James Baker opened the area's first major coal mine operation with the establishment of the Crows Nest Pass Coal Company at Coal Creek, British Columbia. The men had waged a ten-year struggle to have a spur line of the Canadian Pacific Railway built through the area to support their operations. With the completion of the railway, other mines quickly followed: the Erickson and Michel collieries started production in 1899; the Morrissey and Carbonado mines opened in 1902; Hosmer in 1906; and Corbin in 1908. These operations worked a coalfield described by a contemporary geologist as a basin in the shape of "a long-pointed triangle, with its base to the south, the area containing coal being approximately 230 square miles. Its greatest length is about 35 miles north and south and its greatest width about 13 miles." Coal and coke production started small at 8,986 and 361 tons respectively in 1898. By 1906, these numbers had risen sharply to 806,901 tons of coal and 213,295 tons of coke.[4]

Across the Alberta border, the Canadian-American Coal and Coke Company commenced production in 1901 at the Frank mine site, followed two years later by the Western Canadian Coal Company at Bellevue. Western Canadian also established the Greenhill Mine in 1913. A group of American financiers established the International Coal and Coke Company in Coleman in 1903, while the Hillcrest Coal and Coke Company, also an American operation, opened for business in 1905. The McGillivray mine in Coleman commenced operation in 1909.[5] As each new mine opened, the population of the Crowsnest Pass rose sharply with the creation of jobs. The census of 1906 showed 413 residents in Lille, 449 in Blairmore, 915 in Coleman and 1,178 in Frank, while by 1911, Fernie alone had reached a total population of 3,155.

The impressive production and population statistics told one side of the story; equally informative were statistics on mining accidents. Not counting fatalities, 1,435 serious accidents took place in the Crowsnest Pass mines between 1904 and 1928. The common mishaps such as cuts and bruises from coal falls or haulage accidents, broken bones, amputations, lacerations, or burns were not reported in the newspapers. These types of accidents were lost in the normal workflow; they were regarded as the acceptable risks of mining coal. Such injuries were not "news" because their frequent occurrence established them as a fact of coal-mining life. The hidden side to these types of accidents, however, is that more miners were seriously hurt from mine car or haulage accidents, falls of rock, and falls of coal, than the total number of men actually killed in the mine through major disasters (see Charts 4 and 6).[6]

The premier of Alberta, Arthur L. Sifton, summed up the general attitude toward mining after the 1910 Bellevue mine explosion that killed thirty-one men. Sifton stated at the time of the inquest that "he would not allow this mine to be opened until it had been proved to be free of *ordinary* danger"[7] (emphasis added). Similarly, when the Government of Canada's Commission of Conservation ordered a report on the state of mine rescue operations in 1912, the author stated his objective: "to encourage mine-rescue work in Canada by setting forth what is being done in this country and elsewhere, by the establishment of mine-rescue equipment stations, [and] to *lessen* the number of fatalities due to mine fires and explosions"[8] (emphasis added).

This concept of "ordinary" danger, and the idea that fatalities could only be "lessened" but not eradicated, were commonly held beliefs regarding coal mining in the Crowsnest Pass. These reactions could be described as "defensive rationalizations truly believed"[9] on the part of both miners and mine owners. The managers and miners, however, did not adequately understand, or had not previously faced, such dangers as were present in the Crowsnest Pass mines. One geologist significantly understated the problem in 1904 when he noted that "the successful management of a colliery in the West is not always an easy task. This is especially so in the Rocky Mountain coal basins ... with conditions requiring considerable modification [from] Eastern methods."[10]

The dangers that the miners faced were numerous.[11] The coal seams were a significant factor in that they inclined on an average of thirteen to twenty-six degrees.[12] This factor created a need for ingenuity in operating and maintaining the numerous haulage lanes; but multiple haulage accidents still resulted. An abundance of coal chutes employed gravity to move coal down the slope from the mining rooms to the coal cars. Men sustained crushing injuries or death while "bucking coal" (trying to unblock coal stoppages) in these chute lanes. Tom Polidore died in the Western Canadian Collieries mine when "some of the coal upon which [he] was standing stuck in the chute then suddenly [gave] way, taking the unfortunate man ... with it."[13] In addition, steeply inclined seams required a labour intensive operating system; mechanized equipment proved too heavy for the miners to hold up at an angle to the coalface.[14]

Miners used only hand tools and explosives to bring the coal down. Men argued the need for safe explosives in Canada as early as 1892, yet it took a major loss of life before a solution was proposed.[15] Following the 1902 Coal Creek mine explosion that killed 128 men, the British Columbia government commissioned a report on coal mine explosions. The report stated: "The ordinary black powder [gunpowder] which is in use is generally admitted to be dangerous under certain conditions ... we strongly recommend [a permitted list of explosives] be introduced in this province."[16] The government of Great Britain had already established a list of "permitted" explosives, or those explosives deemed safest for use in dry, dusty, or gassy mines. Six years later, the Alberta government finally adopted the British "permitted" list that contained a total of thirty-five types of explosive. Managers had to balance the blasting power of the explosive against its shattering effect. The best explosive would bring down the most coal without the excessive pulverization that made it economically non-viable.[17] "Permitted" explosives did not spark or flame and were therefore safer to use. Unfortunately, "permitted" explosives were more expensive. This created a source of friction between management and the miners, who naturally favoured cheaper explosives; the men paid for any explosives they used from their own wages.[18]

Two other problems compounded the difficulties of mining in the Crowsnest Pass mines: methane gas and coal dust. Coal is formed by the chemical and thermal alteration of organic debris such as plant material. Temperature and pressure increase as the plant material is buried and compacted by overlying sediments. This process, known as "coalification," produces a number of by-products, among which the most abundant are water, methane, carbon dioxide, and nitrogen. As the coalification process reaches maturation, approximately five to seven thousand cubic feet of methane is generated per ton of coal.[19] Since the methane gas is a natural by-product of vegetal decay, the gas is present under pressure at all times during coal-mining.[20] Significant levels of methane gas existed in the Crowsnest Pass coalfields. An analysis of the Michel, Coal Creek, and Corbin mines showed upward of fifteen thousand cubic feet of methane released per ton of coal mined.[21] These extremely dangerous levels were comparable to the few other known gaseous mines elsewhere in the world.[22]

Large quantities of coal dust were also a natural by-product of mining. Horses and coal cars regularly ground the loose coal underfoot into a fine powder. Firing explosive shots released clouds of pulverized coal; pick mining and drilling produced dust as the coal was pounded and broken off into lumps. Many recognized that the dangers from coal dust were higher than from gas alone.[23] Coal dust in even a well ventilated mine had a latent power buried in every grain. If a flammable methane-air mixture (called firedamp)[24] existed in the mine and the gas ignited, the coal dust added fuel to the explosion. The Coal Creek disaster of 1902 is a perfect example of an explosion that spread from room to room, feeding on the coal dust in the air. In his report on the cause on the explosion, W.F. Robertson described the path of the blast:

> ... this initial explosion was conveyed against the air being transmitted by the suspended coal dust which the explosion rendered explosive and which dust was put into the air by the act of

loading; that in each loading place it received augmentation; that the line of greatest force followed such links of augmentation until [it] came to the narrow neck at the entrance ... and here it spread, fan shaped, following the line of least resistance through the old workings.[25]

The men most deeply charred by the explosion were found in the driest and dustiest of the rooms, where the evidence suggested that the fire had burned the fiercest and the longest.

While the actual explosion could produce significant material damage and loss of life, the by-products of the explosion were also deadly. Methane by itself is a non-toxic gas, but if it is present in high enough levels it serves to significantly lower the oxygen content of the air. The result is called blackdamp. The lower the level of oxygen, the more serious the side effects: 17 per cent oxygen in the air results in an increase in the rate and depth of breathing while 15 per cent leads to an accelerated heartbeat and dizziness. If the oxygen falls to between 7 and 13 per cent a miner will experience disorientation, fainting, nausea, headache, blue lips, coma, and convulsions. Anywhere below 6 per cent oxygen will result in eventual death.[26] Even if the gas did not kill a man at first, its effects could influence reasoning and judgment:

> It certainly is a funny sensation the way that gas affects you. The first thing I noticed I was stumbling, and my feet seemed too heavy to lift. Then my knees got shaky ... then there a kind of whirring noise in my ears, and red blobs came before my eyes. There was no pain, but on the other hand it made me feel as though I wanted to laugh ... there was nothing I could say that passed through my mind regarding the seriousness of the situation.[27]

Methane and coal dust explosions require a source of ignition. Many miners worked the seams with naked lights. In the Bellevue mine, there were four separate incidents of gas ignited by open-flame lamps between 6 and 25 August 1906. Arriving home from the mines with singed eyebrows or hair from minor explosions was not an uncommon occurrence. By 1907, the governments on both sides of the border recognized the dangers of open flames and instituted a change from the free-burning oil lamps to the Wolf Safety Lamp. Despite the good intentions behind the change, this created an extensive and prolonged controversy when both miners and managers resisted the process. Among the reasons that the coal companies were reluctant to institute the new lamps was the cost of replacing the old ones. The miners grumbled that the new lamps did not throw adequate light for them to do their work at the coalface. The miners also complained that the safety lamps were deliberately designed so that they could not be relit by the miner if they went out, a measure designed to stop the miners from carrying matches into the mines. If a lamp blew out, a miner would have to leave his place of work and walk out to a lamp relighting station, losing time in the process and risk getting lost in the mine in the dark.[28] On an ironic note, the lamphouse at the Hillcrest mine burned down in 1908, destroying all the safety lamps.[29]

However, there were sources of ignition other than naked flames. Sparks from rolling coal cars or pickaxes could also ignite gas and dust. Blasting, even when done with approved explosives, could still touch off a conflagration, as could an overheated lamp. The controversial finding of the inquest into the 1910 Bellevue mine explosion indicated that the source of ignition was sparks struck by falling rock.[30] The verdict regarding the Coal Creek explosion in 1917 that killed thirty-four men indicated that an extremely unlikely and unfortunate sequence of events touched off the blast. On the day of the explosion, August Leonard and his brother Hector were walking out at the end of their shift, carrying their tools. August appeared to have been carrying a pick, two loose blades for the pick, a shovel, and a saw in addition to his safety lamp. With his hands full of tools, August stumbled in the dark and the pick swung up to puncture the glass of his lamp. Aware that gas had been reported in the mine on eighty of the previous ninety-six days, August anxiously raised the lamp up toward his face to see if the glass had been shattered. By doing so he inadvertently ignited a pocket of gas that led to the explosion.[31]

Legislation and safety precautions did help to ameliorate some of the dangers in the Crowsnest Pass mines. The Alberta Mines *Annual Report* for 1909 stated that "the decrease in the number of fatal accidents for the last two years is largely due to the installation of safety lamps and the use of permitted explosives."[32] In British Columbia, the

Mine Inspectors instituted a test of the standard Wolf safety lamps in 1917 using the new Burrell gas detector. There were some disturbing results. The Wolf safety lamp worked by showing a "gas cap" inside the lamp that indicated the percentage of gas in the air. According to the improved Burrell instrument, men had been routinely working in the mines under unacceptable gaseous conditions, often at more than twice the recommended level. The British Columbia legislation was changed shortly after this discovery, approving a lower acceptable limit of the gas cap showing in the safety lamp. If there was any doubt about the level showing, then safety was to get the benefit of the doubt every time.[33]

However, regulations and precautions could do little to tame nature's unpredictable actions. "Bumps" were random, unpredictable "violent dislocations of the mine workings ... attributed to severe stresses in the rock surrounding the workings."[34] One miner graphically described a bump: "Imagine being in this room and the floor coming up and the ceiling coming down and the walls coming in."[35] Bumps frequently released methane gas in deadly amounts and caused several major incidents in the Crowsnest Pass, including most of the disturbances in the Michel and Morrissey mines and also the 1926 McGillivray mine explosion that killed ten men. "Outbursts" or "blowouts" were sudden explosive discharges from the coalface where a pocket of gas inside the wall yielded to pressure, "blowing out" anywhere from one to thousands of tons of coal. Blowouts were extremely common in the Morrissey mine, leading one of the Chief Inspectors to comment in 1909 that "if this portion of the field is to be worked at all, it must be under those conditions that give the miner some chance for his life. Most of the outbursts in this field have been attended by loss of life."[36]

Coal mining in the Crowsnest Pass was not a complex operation. When stripped to its essentials, the production process of mining could be described as primarily a linear system: "loosen the material, prop up the roof, remove the material."[37] Problems tended to occur when personnel either used faulty equipment or made errors in judgment. Danger naturally increased due to the unpredictability of the underground environment: the "setting can create unexpected, unplanned, and invisible interactions, largely in the matter of the flow of gases and pressures in the complex web of shafts and tunnels and ventilation holes."[38] Some measures could be and were taken by the miners in attempts to control some of the dangers; the use of safety lamps, "permitted" explosives, and also the method of drilling holes in the coalface to relieve the pressure of a possible outburst. The other dangers, including bumps and the sudden release of methane gas, were too erratic and volatile to either monitor or manipulate.

Miners were well aware of the dangers that threatened their safety at work. Residents of a mining town almost always automatically responded to sudden noises with the assumption that something had happened in the mine. When a portion of Turtle Mountain crashed into the valley in 1903, partially burying the town of Frank, Karl Cornelianson "rushed to the door and looked toward the mine, thinking there had been explosion."[39] The Grahams of Frank had a similar reaction. Their "first thought, as always in a coal mining town, was that there had been a explosion in the mine."[40] Due to the nature of their work and their knowledge of the dangerous conditions, miners and their families were obviously primed to expect the worst.

The miners' working backgrounds also indicated that the men were familiar with the dangers of coal mining. The Crows Nest Pass Coal Company imported a total of 209 men from Great Britain in 1907 to their Coal Creek operation. The occupational designation on the employee list clearly states *miner*.[41] An examination of the employee records for this company also indicates that workers applying for positions had previous experience. A random sampling of 515 European workers employed from 1907 to 1909 reveals a total of 2,264 years of experience, or an average of 4.5 years per man. This experience could range anywhere from less than a year in the case of John Sturko, an Austrian timberman, to the forty years of Frank Licker, a German contract miner. Among the other employees with mining experience were men from the United States: W.J. Oliver from Buxton, Iowa with eighteen years' experience; Art Neros with twelve years in the Washington mines and James Marnagu with fifteen years' experience in the Montana mines.[42]

Young boys grew up in the Crowsnest Pass equating mining with danger. Experience taught them that mining frequently resulted in death. During a trip to Lille in 1905, John Gray noticed a number of men going in and out of a shack: "I was curious so I went in too. What a shock I had! There I saw my first dead man. He had been killed in the mine." Boys sometimes learned about the dangers of mining firsthand during play. Frank Horejsi and his friend Frank Cizek went exploring in an old mine at Frank. Although at sixteen they should have known better, they lit a match in the mine to see their way. The gas ignited with the result that both boys were burned severely. Despite this experience, both later became miners; Cizek would die in an accident in the Blairmore mine in 1934.[43]

Most of the boys growing up in the Pass absorbed mining knowledge by being part of a mining family. The majority accepted that they would follow their fathers into the mines. Erwin Spievak, a miner from Michel, remembered: "Coal mining was more or less for family. If your father was a miner, you'd go down to the office every morning before they started and look for a job."[44] Boys would listen to their fathers "talk pit" with their friends around the house or in the bars during their off hours. The Goodwins of Coleman remembered: "Children became accustomed to talks, debates and arguments about such things as 'levels, pillars, slopes, chutes' and all other aspects of the work."[45]

Sons also accompanied their fathers to the mine and hung around the area absorbing the sights and sounds. Sometimes this had tragic results: William Harvey, eleven years old, "met with a fatal accident while playing along with three other boys in the yard of Mine #87 operated by West Canadian Collieries."[46] The cemeteries of the Pass also indicate that mining was passed down from father to son, and that dying in the mines could be a family affair. Thomas Phillips died in the Michel mine explosion during the tremendous thunderstorm of 1916, followed sixteen years later by his son John Thomas; William Puckey was among the thirty-four dead in the 1917 Coal Creek explosion, while his son Thomas died from a haulage accident in the same mine seven years later. The Munkwitz family also lost both father and son to the Michel mines between 1910 and 1921.[47]

To be accurate, however, not all men hired as miners had previous experience, came from mining families, or were familiar with the dangers. General Manager William Blakemore of the Crows Nest Pass Coal Company wrote in his *1899 Annual Report*:

> We have had no satisfaction from the Western [British] men whom we have employed. They have been a source of constant trouble and expense to us, being as a rule inexperienced in Coal Mining and accustomed to high wages, and short periods of employment ... it is my intention to employ as few of these as possible.[48]

The Company obviously followed Blakemore's suggestion not to hire British workers; the vast majority of miners killed in the 1902 Coal Creek mine explosion were of Slavic origin. An examination of the Crows Nest Pass Coal Company's employee records also indicates that hiring practices were sometimes far from selective. The Company seemed willing to hire anyone regardless of previous or even relevant experience: Fred Dowd (French) was hired in February 1907 after a previous career as a bed maker in Spokane. Tony Gross (Galician) applied in September 1907 after two years' experience as a shoemaker in Fernie. Bernard Savage (English) proved perhaps the oddest hire. The Company employed him in March 1907 despite his only listed experience as a librarian. Savage quit in April, was hired back again in June and then finally discharged.[49]

All greenhorn miners soon acquired the basic facts about the dangers of mining. Yuriy Frolak, a recent immigrant from the Ukraine, had no previous experience in coal mining when he started at Hillcrest. Despite this deficiency, he quickly learned about the hazards inherent in his new job. He described the system of teamwork vitally necessary to working safely in the mines:

> A high degree of cooperation not only between the two-man teams, but all miners, was essential, and any breach resulted in instant firing. The ever-present danger of being buried by cave-ins or being crushed by falling coal, beams, or coal carts, or being suffocated, made everyone aware that survival could only be achieved by co-operation and mutual assistance, and all were dependent on each other if the hazards were to be minimized.[50]

Inexperienced miners such as Yuriy were usually paired with a more experienced man. This ensured that knowledge about the mining methods and the mine itself was passed on in a secure fashion.[51] Frequently these two-man teams were composed of father and son as the knowledge was handed down through the generations.[52] However, there were also specific jobs that required an experienced man. In the case of drawing pillars (removing the supports from the rooms from which the coal had been mined) "a great deal depends on the miner. Unless a man is experienced with this class of work ... it is not advisable to assign him to the duty except in company with an experienced pillar-miner."[53]

Work attitude is an important factor in determining how a person responds to the pressures of the job. Some miners did enjoy their work. Leslie Brown from Hosmer described "those who liked the job, even enjoyed the routine:"

> ... they had a feel for the depths, for the dim lights in the dark, the brush of air flowing in ventilation, the movement and softened noise of cars, the satisfying roar and crash of coal tumbling down the chutes in response to a man's urging.[54]

Charlie Drain, later elected as Member of the Legislative Assembly for Pincher Creek-Crowsnest, remembered: "I left [the mines] at 22 with great regrets. I had grown to love the companionship of the people who worked there, the smell of coaldust ... the whole thing."[55] Albert Goodwin recalled mining as "a wonderful life." He especially enjoyed the process of actually digging coal, describing it as the "finest job of all."[56] A common theme emerges among miners who extolled the virtues of coal mining. Although they were initially afraid of the mine, they got used to the work and would not have traded the life for anything. Jack Marconi of Coleman "stayed underground 43 years and I didn't mind it except when I was 15 years old,"[57] while Gino Busato of Fernie remembered that "once you got used to it, it's just like we're sitting here [at home]."[58]

However, for every miner who enjoyed his work despite the dangers, there were many others who did not. Whether due to the unsettling experience of working underground, the heavy workload, or the difficult conditions, many miners never became used to digging coal. Alrik Tiberg stated that there were days when he went into the mine and watched the daylight recede from view, never knowing if he would see the sun again.[59] Ferdinand Fabro echoed these sentiments when he remembered "spending many dangerous and back-breaking hours with little pay while employed at the International and McGillivray Mines."[60] Yuriy Frolak described his own reaction to his first days of mining: "The traumatic experiences of the seven boys [recently arrived from the Ukraine] in changing from farmers ... to miners digging out coal from the bowels of the Rocky Mountains ... are best left to the imagination...."[61]

Why then did the miners stay in a job fraught with danger? If they disliked the work and the conditions, why didn't they move on? One standard answer points to the limitations on job mobility during this period. Many miners were recent immigrants, drawn to the area by the promise of steady work at decent pay. Often they had left behind the turmoil of their native land and a lack of prospects for themselves and their families. After spending all their money travelling to the Crowsnest Pass, the expense of moving again and the uncertainty of finding work elsewhere kept many immigrants in the area. Pasquale Perri of Fernie stayed because "it was too good of a job to quit ... other people couldn't find no work and I had one I keep it [sic]."[62] There were essentially only three avenues of endeavour for recent immigrants with little or no experience or specific trade: homesteading, working on the railroad, and coal mining. Immigrants often found coal mining more attractive as it offered higher wages but did not require the men to travel away from home for long periods of time.[63]

Limited job mobility cannot explain all the actions of the miners. Regardless of their nationality, workers continually moved both into and out of the mines, especially in the early years of the twentieth century. When the Frank Slide buried the entrance to the mine in 1903:

> No one had any idea as to how many people were missing as men were coming and going all the time and the coal companies did not even take down a man's name as most foremen could not read or write and the men spoke no English and most them had names that no-one could pronounce or write anyway.[64]

This observation is patently obvious for the British Columbia mines as well. When accidents were reported in the papers, frequently the names of the miners were not given, and only sometimes their nationality, suggesting that these were single men with no ties whom no one could put a name to.[65] Some of this movement can be explained by the unpredictable demand for coal, layoffs or strike actions. However, it could also indicate men leaving for other jobs, perhaps in lumber camps, other mines, other towns or with the Canadian Pacific Railway Company. The employee records for the Crows Nest Pass Coal Company list many men as young and single. They had no ties in town in terms of family, no obligations to feed wives and children, or create a stable environment for them. These men were free to move between towns and occupations before they either put down roots or settled into married life.

Many miners held jobs in the mines as a means to an end. Polish immigrants in particular shared the common goal of owning land in order to achieve economic independence. When Paul Pieronek arrived in the Pass, he intended "to make enough money working in the coal mines to return to Poland and buy more land and a house there." Other Polish workers also planned to work just long enough to buy land for farming, but still expected to stay in the area. Many miners had families who worked on the land while the men laboured in the mines for extra wages. These men may have taken some time to realize their goal (Karol Rosmus worked for twenty-four years in the mines before moving permanently to his farm), but many Poles regarded mining as a "temporary" occupation.[66]

Some erstwhile miners only worked for short periods in the mines and then moved on. Ivy Haile of Fernie remembered the "men used to come in after the harvest on the prairies. They would work in the mines for awhile. A lot came from the East Coast, some stayed, some made enough money to go back to New Brunswick or Nova Scotia."[67] Joe Fabin testified at the 1907 Alberta Coal Commission inquiry that he had worked in eight different mines in the United States and Canada over a period of some eleven years, "always looking for a better show."[68] He had settled in Frank in 1905, reluctant to move again on account of his family. Other miners were simply unsuited for the job. Jim Cousins "got a job at McGillivray but after a brief, inglorious mining career ... went back to school in September."[69] Harry Gate worked for a while bucking coal but eventually became dissatisfied. He moved on to work at the Red and White Store in Coleman, then the Grand Union Hotel and eventually became a forest ranger.[70]

Others who tried mining decided to quit sooner or later because of the dangerous conditions. Perhaps the most famous example was Emilio Picariello, who worked in the Coal Creek mines for a few shifts. Deciding eventually that there must be an easier way to support his family, Picariello left the mines to become the entrepreneurial "Emperor Pic" on the Alberta side of the border.[71] Two members of the Dugdale family "worked in one of the Coleman mines for one day, but decided it was too dusty."[72] For some, it took a major loss of life to induce them to move on. After the 1902 Coal Creek explosion, "many of the miners ... declare[d] in no modified terms their intention of quitting the mine just as soon as all attempts to recover the bodies have been abandoned ... the men state[d] that they do not further care to risk their lives."[73] A personal tragedy convinced others to find another location or position. The McKinnon family from Glace Bay, Nova Scotia left the Pass after two brothers were killed in the 1914 Hillcrest mine explosion.[74]

However, the majority of men stayed in the mines. While some miners without doubt did not have adequate resources or opportunity to move on, modern studies suggest that a *perceived* lack of job mobility may have kept some of the men in their jobs. "Learned helplessness" develops when workers either think or know that the prospects of getting another job are limited. This outlook could give workers a sense that they are not in charge of their destinies and that events in their lives are not controlled by their own actions.[75]

There are four factors that might increase occupational accident fatalism: (1) the rate of previous occupational accidents; (2) perceived risk of personal injury; (3) ability to find other work; and (4) a desire to leave or stay. Workers with high accident rates and dangerous occupations tend to have a correspondingly high level of accident fatalism. These workers are more likely to resign themselves to the inevitability of personal injury accidents. Workers who felt themselves trapped in their jobs had the worst accident histories and the highest perceived risk of injury. This suggests a correlation between a worker's attitude toward his perceived lack of mobility and an increase in a fatalistic attitude toward safety on the job.[76]

Miners whose jobs directly involved them in the basic mining processes of "drilling, blasting and mucking" were asked questions related to understanding how employment under dangerous and stressful conditions would affect their attitudes. The results indicated "individuals who believed they were 'stuck' in one job experienced a sense of hopelessness."[77] The attitude in town after the 1914 Hillcrest mine explosion was indicative of this feeling: "They just kept on, and most all of the men went back into the mine. They didn't have much choice."[78] Historically, this attitude has not changed. Workers at the Westray Mine in Nova Scotia in 1992 who knew of the dangers were also reluctant to move. Shaun Comish remembered: "I still can't figure out why I didn't quit a dozen times ... it is hard to explain to people how scary it is to leave a profession you have worked at for twelve years and look for work elsewhere."[79]

Previous analyses of miners' responses to danger have tended to centre on their apparent disregard for the consequences of their behaviour in dangerous situations. The high accident rate in the Crowsnest Pass appears to be linked to a calculated lack of safety consciousness on the part of the miners: "... ignorance, neglect and carelessness of those who worked, inspected and owned the mines ultimately created the conditions in which avoidable accidents needlessly claimed victims."[80] This conclusion came after looking at the situation from an essentially administrative point of view; the interaction between management, government, and the miners indicated that they made choices to deal with danger and yet still retain the economical viability of mining. Individual miners' responses were not examined.

In contrast to this point of view, miners have been interviewed to investigate their reactions to dangerous situations to "better understand the reasoning, thought processes, behavioral reinforcement, or other possible reasons for unsafe behavior." The interviews uncovered six types of responses that miners employed when dealing with dangerous conditions. Only one of these responses could be identified as a lack of safety consciousness or "risk-taking" behaviour (34 per cent). The other five were "unconscious behavior" (26 per cent), "unknowing behavior" (18 per cent), "assumptions" (9 per cent), "conscious choices" (7 per cent) and "following orders" (6 per cent).[81] Although these findings come from modern-day miners, numerous examples of these types of behaviours can easily be seen in the Crowsnest Pass in the early part of the twentieth century.

Admittedly, the higher number of examples relating to "risk-taking" behaviour indicated this response should be placed at the top of the list among Crowsnest Pass workers as well. The Alberta District Inspector of Mines concluded that "[Wajeich] Konopaski was to blame for an accident [in the McGillivray mine because] he should not have worked underneath the top coal knowing that there were two loose sides to it."[82] Steve Billesky died when he had "gone on mining the coal from underneath the slab of rock ... another case of a miner taking a chance and knowing it, contrary to advice by officials of the mine."[83] Joseph Klis had a fatal accident when he and his brother neglected to put up enough timber, despite knowing when and how to do so. Martin Klis told the inquiry that he and his partners had wanted to load another car with coal first.[84]

"Unknowing behavior" occurs when a person is unaware of danger at the "action moment," or the moment in time when the decision was made to perform the action. Gustav Devignut entered the Bellevue mine without permission while it was being inspected. He pleaded, through an interpreter, ignorance of the law.[85] Peter Sadleish fractured his leg at Coal Creek while unloading timber, "not noticing that another man was already unloading [and] was caught by a stick of timber from the car."[86] Men who had followed all the rules and put in place all the safety practices made "conscious choices." When "coming off shift, [Peter Murray] got on the footwalk to keep clear of the horses, slipped and sprained his ankle."[87] Thomas Hynd testified at an inquest: "Burns [his partner] examined his side and I examined mine, and Burns came over and examined mine."[88] After agreeing it was safe, they started to knock out the pillar. The resulting coal fall killed Joseph Burns. Both miners had over thirty years of mining experience.

Miners who said "their minds were elsewhere" characterize "unconscious choices." A series of accidents at Coal Creek between 1906 and 1907 indicates that miners clearly had other things on their minds as they were coming off shift: Richard Higham severely scalded his right foot by stepping in a bucket of hot water on his way to the washhouse. Edward Sherwin was also injured when he stepped in a hole filled with hot water while on his way to change clothes and head for home.[89] D. Vilityk died in the Frank mine in 1907 after wandering into a section of the mine filled with gas. Although Vilityk had worked twelve shifts at Frank, and had been back and forth through the crosscuts as many times, his mind must have wandered at a crucial moment causing him to lose his way.[90]

"Assumption" occurred when a miner assumed all safety precautions had been taken and that their own actions would not cause an accident. Fred Kubalo was crushed at Michel when the switch boy gave a signal bringing loaded mine cars down the mine track by mistake.[91] John Mazur must have assumed that the doors were working fine the day he died in 1917 at the McGillivray mine. While bringing out a load of three cars, the noise of the mine could have prevented him from hearing the shouted warning from the door trapper that the doors were stuck. Mazur "continued going ahead and crashed through the door."[92]

"Following orders" describes performing tasks as directed by a supervisor, even when confronted with a dangerous situation. At Bellevue, supervisors directed one shift to continue digging coal in an area despite warnings from an earlier shiftboss that the face would not stand the pressure. The wall collapsed, flooded the mine and killed two men.[93] Joe Boline received a broken nose, cuts, and bruises at Michel in 1908. Although he had waited for the signal before setting his trip of mine cars in motion, the signal was premature.[94] Tomaso Mazza died at Coal Creek in 1913 when his supervisor told him the rope he was using for his mine trip was strong enough to hold the weight. The rope broke, and the mine cars ran over Mazza, fracturing his skull. William Wilson, the supervisor, eventually received a four-month suspension for his role in Mazza's death.[95]

These examples show that not all miners responded to hazards with foolhardiness or a casual attitude. There are as many accounts of miners taking risks and chances as there are of miners obviously cognizant of the dangers who made attempts to avoid injury or death. The death of Matthew Turnbull in the Coal Creek mine in 1920 illustrates both of these behaviours. Turnbull had received adequate training and certification to reach his current station as a fireboss. At the inquest into Turnbull's death, Mike Koziec explained through an interpreter that he and his partner were following the orders of their overman in digging holes to timber their area and prop up the roof. Turnbull arrived at their working place and immediately directed them instead to place the props in another spot. Completely disregarding the protests of both Koziec and his partner, who both knew that the roof might cave if the props were changed, Turnbull proceeded to withdraw the timbers. The roof caved in and buried him. Although the inquest verdict declared Turnbull's death an accident, the *Annual Report* described his behaviour as "reckless" and made a point of stating that the miners in the room had protested against his actions.[96]

Negative reinforcement for unsafe behaviour was also a powerful deterrent to miners, either in the guise of a near miss for themselves or by the injury or death of a partner or family member. Mike Koziec had survived eight years as a miner before he witnessed the death of Matthew Turnbull. When Koziec himself died in the Coleman mine eight years later, he was given full accolades as a "careful miner" who died solely through lack of first aid supplies.[97] Other miners took living through a near miss or witnessing a fatality as a cue not to redefine their own behaviour, but rather to quit the mines entirely. Gino Bursato knew a number of men who would not go back into the Coal Creek mine after a bump sealed them in for a brief period.[98] Filippo Rogolino never returned to the mines after an explosion occurred the night he was scheduled to go on shift.[99] However, negative reinforcement could only go so far to influence men: "... allowing an employee to experience an injury, illness or fatality is not an option for preventing future unsafe acts and developing safe work behavior."[100]

Unsafe behaviour in miners could also have been a learned response from messages sent directly or indirectly by management. The pressure for production might lead management to praise or reward miners despite unsafe behaviour:

> There exists a conflict of values when an employee values praise over performing their work safely. In some instances, a supervisor may know an activity is unsafe, yet look the other way when the work is being performed. This supervisor and, perhaps, the employee both place a greater importance on working quickly versus working safely. Safety policies must not be violated to accommodate production crises, which tells the employee that production is valued over safety.[101]

Miners and management in the Crowsnest Pass displayed equal evidence of this type of learned behaviour. Hugh Dixon, a miner with forty years' experience, testified at the inquest into the 1902 Coal Creek explosion:

> Q – And you knew there was too much dust?
> A – These men [supervisors] seen this thing as well as me
> Q – Did you speak to them about it?
> A – Why should I call their attention to it if they could see it for themselves?[102]

Although aware of the dangerous accumulations of dust, Dixon responded by adopting management's apparent lack of interest. A decade later saw no change in the nature of this response. At the inquest into a 1915 Coal Creek explosion, David Shanks the fireboss testified:

> Q – Do you consider that your practice of discipline as mining boss was according to good mining practice?
> A – It was according to orders we received and the way we had been treated for the last six months.[103]

Shanks believed, from the information he had received, that any safety measures or instructions he had given to the miners were in full accordance with those of management. The Bellevue mine provided another example. Management painted a "Safety First" sign on the concrete buttress facing the miners as they headed into the mine every morning. Yet, as noted previously, management directed a shift to continue mining a coalface with fatal results.[104]

Exhibiting safe behaviour may only be a successful response to danger among miners when safety becomes a personal value. Some of the Crowsnest Pass coal companies created lists of *Special Rules* meant to govern the men's behaviour and to reinforce common safety precautions.[105] These *Rules* clearly stated such simple safety measures as not riding on empty coal cars, mining "only in safe areas," or not smoking or carrying matches. However such rules were essentially ineffectual without the reinforcement needed to establish these rules as a personal value. Simply posting mining regulations and expecting miners to read and follow them, without emphasizing their importance with significant penalties, only provided loopholes and ambiguity for the miners.

As an example, attempts to stop the miners from carrying matches into the mines tended to be inadequate and sporadic. Although fines existed, only periodic searches were done. The mine managers wanted the provincial governments to enforce this simple precaution with legislation; the government laid it back at management's door, suggesting they control the problem internally.[106] By 1908 the British Columbia government was complaining that offences against the rules were becoming all too frequent. The provincial Inspector suggested that instead of fines, a miner's certificate would be revoked. This measure would effectively demote a miner from contract status and would significantly affect his pay. Whether or not this measure was either taken, or was at all effective, can perhaps be measured from the comments made at the inquest into the 1915 Coal Creek explosion. When the examiner questioned Bernard Caulfield, the Mine Supervisor, regarding the discipline in the mine, Caulfield replied:

A – It would not be very high.
Q – You would not consider that there had been that much discipline there?
A – In my opinion, there was no discipline over the holiday time.
Q – It was a sort of haphazard, happy-go-lucky state of things?
A – That is how it turned out.
Q – In fact only God was looking after the conditions that existed in that mine at that time?
A – Yes, that is how it appears.[107]

Among the pantheon of miners' responses to the dangers of their work, their fatalistic attitude is often remarked upon, both in their own comments and by outside observers. Fatalism is a belief that all that happens is predetermined and therefore inevitable, while a fatalist is one who accepts and submits to what happens, regarding it as inevitable. Gino Busato of Fernie escaped several close calls with injury or death. When asked if he went back into the mine after being trapped following a large bump, Gino stated: "Oh yeah. Why not? If it's your time you're gonna go whether you're in or out [of the mines] it don't matter ..."[108] This attitude is akin to making a deal with death.

The miner accepts the dangers of his occupation in return for a good wage and a chance to make a living. He may never know when death might choose to claim him, whether in the next hour, day, month, or year, but he accepts the terms demanded of him.[109]

A miner's adoption of this attitude is easy to understand. There were too many examples demonstrating the fickleness of fate that arbitrarily either took or spared lives. Even though Thomas Hynd and Joseph Burns worked at the same coalface at the same time, doing the same job, the coal fell one way and only killed Burns.[110] Bobbie Jenkins worked in the McGillivray mine and recalled "leaving his partner at work for a few minutes to find out the time."[111] He returned to find his partner dead, killed by a coal fall. Some men found themselves spared from certain death on one occasion, only to be killed on another. In 1903, Charles Elick had been one of the sixteen miners trapped underground in the Frank mine following the Turtle Mountain slide. Undeterred by this experience, Elick went straight back to the mines, only to be killed in 1914 at the Hillcrest mine.[112] James Maddison survived the Hillcrest explosion of 1914, fractured his leg in one accident and in another was buried up to his neck in coal by a coalfall. Maddision "marvelled that he escaped death."[113]

Adoption of a fatalistic attitude would have been an extremely useful defensive mechanism. Fatalism provided a rationalization for all the apparently random injuries and deaths. It also shielded the miners from the thought of what *might* happen. Handing over the possibility of death to fate would take much of that responsibility and worry away from themselves. This is not to suggest, however, that fatalism desensitized miners to their surroundings or made them feel invulnerable. Following a blowout at Coal Creek in 1917, the miners responded with a "general panic" and "lots of comment" due to the fact that a similar event had occurred a year earlier in the same mine, killing one miner and imprisoning several others.[114] Rather, fatalism provided a thin layer of insulation that protected them from the continual pressure of the very real possibility of injury or death.[115] One miner summed up this sense of fatalism: "I'll live as long as I'll live, and I'll die when I die. I have no say in it. *But I don't go out of my way to push it."*[116] (Emphasis added.)

Fatalism also served to relegate fear to the background. "You get used to it" was a common refrain heard from miners. Tony Cornil described his experience starting work in the Frank Mine:

> I had never been in a mine before and here I was with three other fellows going down, down deep into a dark hole with only a little lamp to give us light. It gave me a strange sensation – maybe that I would never see day light again – but it's strange how one can adapt oneself and after a few days I didn't mind it anymore.[117]

Miners may seldom have verbalized their fears, but they definitely existed below the surface somewhere. Albert Goodwin, who always enjoyed being a coal miner, nevertheless noted: "Although the feelings were rarely expressed, the overhanging threat of individual accidents or group disasters was ever present."[118] Miners channelled this fear through the filter of fatalism. Awareness of danger and acceptance of the possible consequences afforded the miners at least a limited sense of freedom from the oppression of continually living in fear. Fatalism also enabled miners to acknowledge the dangers and their own feelings by couching them in such a way that they celebrated the fact of their own survival.[119]

Where and how did this sense of fatalism develop? One possibility may be that this was a learned social response, handed down through the generations. Women were an important factor in this response. Their calm demeanour and (if only outward) acceptance of their husband's occupation "taught children to accept the conditions of their father's employment."[120] In the Crowsnest Pass, wives were often laconic when asked how they felt about their husband's jobs. Kay Yates stated simply: "There was an explosion at the McGillvary [sic] mine when he was there and for a short time, I thought he had been killed. But he was able to get out."[121] Living in Coal Creek meant that Grace Dvorak often experienced the sensation of the mine moving underneath her: "... you just had to get used to the bumps. Maybe the dishes would rattle in the house, or the pictures would fall off the wall."[122] Other women accepted the risks because of the opportunities involved in having a miner for a husband. The Dunlops returned to Coleman with their family from a farm in Saskatchewan "to provide their children with educational opportunities."[123] Another miner's wife unsuccessfully campaigned for her husband to stay in the mine because moving to a farm was "too far away from medical help and civilization."[124]

Other miners accepted the dangers as part of a mining tradition and were more than willing to follow their fathers into the mines. Dale Montalbetti of Coleman was a third-generation coal miner. The Goodwin clan, grandfather, father, and four sons, "had been miners in England and in Nova Scotia and they saw no reason to change their trade when they came to western Canada. As the male children grew to manhood, they expected to follow their fathers into the pits."[125] All four of the Murrays followed their father into the Hillcrest mine, only to be killed in the 1914 explosion. From the list of thirty-one dead at the 1910 Bellevue explosion, at least four pairs could have been brothers and another pair father and son.[126]

As is always the case when discussing miners' responses to danger, there were exceptions to the rule. Some women, especially, were not as accepting of the dangers their husbands or sons faced. Joseph Gigliotti from Michel had heard his mother crying in the night on many occasions as she admonished his father to quit the mine.[127] Dale Montalbetti remembered: "My mother had six brothers who were killed in explosions in Welsh mines. She said she'd skin me alive if I ever went into the mine."[128] Gino Remo attributed his mother's grey hair to the fact that he, his brother, and his father were all miners. The men all worked on different shifts so that for every hour of every day, there was always one of her family underground.[129] One miner's wife commented, "I'd rather have a little less money than see him go down that hole. You go down so deep and I couldn't stand seeing him come home being so tired." She worked herself for wages until they could both leave and set themselves up as farmers.[130]

Some fathers, although content to work in the mines themselves, actively discouraged their sons from following in their footsteps. Alrik Tiberg "would not encourage a son (if he had one) to work in the mines,"[131] a sentiment that is also echoed in the familiar mining song "Don't Go Below":

> Your gran'dad was a miner and he raised me like himself.
> A poor boy at the age of twelve put toys upon the shelf,
> But I broke a leg at fifteen, and an arm at twenty-two,

> A thousand cuts and bruises, son; I don't want that for you.
> So don't go below where your daddy had to go.
> Study hard and learn your lessons well,
> For you're momma's only boy
> And you're daddy's pride and joy,
> And I'll tell you straight, son
> Don't go below.[132]

Perhaps Joseph Gigliotti's mother had some effect upon his father; the elder Gigliotti flatly stated that he would refuse to talk to his son if Joseph followed him into the mine. Of the Guzzi family from Fernie, only one brother entered the mine, the rest choosing other work in town.[133] Ferucio DeCecco remembered that he had the opportunity to work in the mines only because his father had died: "... if my father had been alive, he would have broken both my legs before he let me go underground."[134]

Mining songs can also serve to graphically illustrate the miner's sense of fatalism. One song in particular is instructively entitled "The Miner's Doom," whose last stanza concludes with the words:

> The widow lamented the fate of her husband,
> Broken-hearted she cried on her dear loved one's tomb:
> "To the world there is left three poor little children,
> Whose father had met with a coal-miner's doom."[135]

This song celebrated the best that a miner could be by describing the doomed miner as a cheerful family man "loved by his friends and his neighbors and all." He was willing to "work like a hero and face every danger" in order to provide for his wife and children. These attributes describe an idealized version of a miner, who nevertheless still met with the "miner's doom." The song may have provided some comfort to actual miners who would also have found an outlet in the song for their own real fears and emotions.

All of the responses noted so far are essentially passive. However, other factors need to be taken into consideration. Miners also reacted in an active and positive manner when they chose to value other aspects of the mining life over and above the dangers they faced on a daily basis. These factors included a strong sense of warmth and camaraderie fostered between miners and the creation of a close-knit community.

Miners also celebrated the challenges of their work in a positive way by reacting with a fierce independence and pride. Miners were above all practical men. Although on one level they may have accepted the dangers of their workplace, they also worked through the unions, one of the few channels available to them, to try to change their working conditions.

The sense of pride fostered by mine workers had much to do with how fathers handed on the work ethic, training, and education of mining to their sons. Mining songs memorialize the tradition of handing down mining lore from father to son:

> My father bein' a pro miner, he knew his way around,
> His guidance bein' the greatest help in the dangers all around.
> The danger's so prevailin' that no one ever knows
> What time a prop or boom might snap, and you may never see your home.[136]

A boy's apprenticeship started the day he was born into a mining community and continued while he was growing up and listening to his father "talk pit." This type of discourse could fire a young boy's imagination as to his father's work. Victor Bielek of Hillcrest described the talk: "You know, a coal miner is just like a fisherman with his fish tales. In the bar, we dig more coal than in the mine, because everybody brags about what they do."[137] When a son followed his father into the mine, he stepped into a well-defined role in the community that his elders and his peers understood and respected.[138] One miner remembered "naturally I wanted to go into the mine."[139] He did not question the role that had been prepared for him and that he accepted as his birthright.

When boys first entered the mines they usually started at the bottom rung as door boy, then a mule driver and finally a miner. Door boys were entrusted with the vital task of regulating the flow of air into the mine. Many young door boys faced their first true test as a miner alone: "... the darkness, stillness, and loneliness combined to oppress a boy's mind making him serious far beyond his years. [This environment] bred in him a sense of responsibility and taught him the lesson of courage."[140] From door boy, a young man would graduate to being a driver, a job that took him into all the sections of the mine where he could absorb knowledge by watching the miners at work. Although boys did not learn any direct technical skills as either a door boy or driver, they were becoming "pit-hardened" and acquired a basic understanding about success and survival in the mines.[141] When boys were old enough and strong enough, they underwent the initiation of a true miner at the coalface. Boys usually started work with their fathers as the junior partner loading coal. The more experienced man would bring the coal down, teaching through example how to dig out the greatest amount of coal from the face.[142] Through working with their fathers, boys experienced a sense of satisfaction "that [they were] learning to do a man's job."[143]

Fierce pride in their craft and independence of spirit sustained miners throughout their working lives. Mining was a challenging occupation, despite its sometimes monotonous tasks. Especially for contract miners, each coalface presented a different challenge in the way the coal was shaped and angled. Using a pick meant not so much brute force as finesse to bring the maximum amount of coal down in the fastest time.[144] Judging the fall of coal, placing the shots in the correct place, timbering the area accurately: all of these tasks required a knowledge and skill that could only be developed through experience and patience.

Miners were craftsmen who worked in a "workshop in which the constantly varying materials presented by nature [were] processed in a changing and threatening environment."[145] John Fry of Blairmore fondly remembered that: "in coal mining, every day is different; that's what I like about it. You can't say that it's boring."[146] Erwin Spievak mined in Michel for forty-five years: "... mining coal as a contract miner was good work. There was always something about it that was a challenge."[147] Despite all the hardships, men took satisfaction in their work because of their skill. Peter Aschacher of Blairmore "saw many changes in mining, many improvements, many accidents, many strikes and many hard times, but [I] enjoyed the work because [I] was one of the best."[148]

Many of the miners in the Crowsnest Pass were either new immigrants or the first-generation descendents of European immigrants. Their work values were heavily based on a sense of realism in that their lives were preoccupied by survival strategies and family welfare. These early immigrants "socialized their children to accept, and even to pursue, steady industrial jobs." The immigrants derived "a system of meaning – a source of satisfaction" from their work in the mines. They developed a sense of pride not only from their physical skill but also from working hard and providing for their families.[149] Many of the oral histories and accounts from the Crowsnest Pass in the first part of the twentieth century focus on the need to provide for family, the harsh realities of early life tempered by the acceptance of these realities and attempts to alleviate them through fun and family. There was the sense of pride in their work, in braving the dangers, in providing for family and in sticking it out through good times and bad. When the men of the Crowsnest Pass looked back on the early days of mining, they lamented the fact that present-day miners didn't seem to have the same sense of work ethic or pride as they used to: "I wouldn't want to work with the guys who were there when I retired. All those guys wanted was their paycheque. They didn't give a damn, and they couldn't work like the old miners did."[150] Although this could be dismissed as simple nostalgia on the part of one miner, the same sentiments are echoed by many others.

The camaraderie of miners deserves to be noted as another important active response. This appears to be a universal element found in all mining communities throughout history. The judge at the Westray Mine inquiry noted in 1992: "… the industry is very close-knit with an interdependence, camaraderie and fellowship that may be unique in modern-day business."[151] As expressed by a member of the Goodwins, one of the larger coal-mining families in the Crowsnest Pass: "Like all the other [miners], [we] suffered when fate struck one or more of the mining fraternity, whether it was a relative, a close friend, or a distant acquaintance."[152] This sense of community could envelop a man soon after his joining the mine. William Powell died after only ten months as a miner, yet an anonymous "Bellevue friend" thought highly enough of him to print a tribute to his passing in the paper: "To know him was to love him."[153] The mining fellowship could also help the men through hard economic times: "When the work was really scarce, the men would gather in silent clusters to wait, probably gaining a little courage from each other's presence."[154]

Mining songs celebrated the camaraderie of the mining community in a variety of ways which could include the simple pleasure of eating lunch with a partner or friend:

> When you're just about fagged, and you've sure lost your punch
> What joy to relax while you have a wee lunch.
> You joke with the rest and you laugh while you dine
> With your battered old piece-can down deep in the mine.[155]

Songs also commemorated the type of men found in the mines, tending to portray them as a "cut above the rest": "There was a sense of comradeship that made you feel at home; you would never meet with better men if through the world you roamed." Mining songs also described the strong sense of community or of helping a brother in need: "Whenever a comrade struck bad luck, whatever his race or creed, His buddies were always ready to help in his need."[156]

Above all, this strong sense of fellowship provided a vital function for men facing danger every day. Working within a "brotherhood" bestowed a crucial support system on miners, both in their everyday working lives, and also as a means of survival in the case of an accident or disaster. During the inquiry into the McGillivray explosion of 1926 this "spirit of camaraderie that exists among mine workers" became very evident:

> When quick action is necessary to save lives, risks are taken which would be frowned very severely upon under ordinary circumstances, but if there is a chance of saving a life, the men knowingly risk [their] own in order to try and rescue a comrade.[157]

The miners' code, unwritten but understood by all who worked the mines, could be described briefly: "... because they worked together they had to take a certain responsibility for the other's safety."[158] Part of this code meant that miners always watched out for each other;[159] another part supported the notion that a man should conduct himself in such a way as not to make conditions worse for his neighbour.[160] In the case of an accident or any dangerous situation, miners knew they could count on their partners in the same room or in the same mine to do all they could to make sure everyone got out safely.

The miners' code, while certainly a positive aspect of a man's working life, should not be emphasized beyond what it actually embodied. Miners were not saints but eminently practical men. They faced a common danger and by adhering to the code, every miner knew that if they pulled a fellow from trouble on Monday, then on Tuesday, or in the next week or year, it could be themselves that needed help. The code did not submerge the miners' personalities by making them more altruistic, compassionate, or less combative then their characters allowed. It could not ultimately save their lives, or make them stronger than they were individually, or even make one man form a friendship with another if he was not so inclined. Instead, the code could serve as a basic understanding among men with different languages, cultures, training, and experience: "... when a man gets hurt, all come to do what they can. It is an unwritten law...."[161]

Camaraderie among miners could take another form as exemplified by membership in a mining union. The unions provided the miners with a voice, the solidarity and bargaining power that they could utilize to try to advance their own safety agendas. The Coleman Miners Union Local 2633 of the United Mine Workers of America passed several resolutions in 1908, trying to draw attention to the common dangers in the Crowsnest Pass:

> Whereas, the vocation of coal mining is known to be hazardous calling [sic] ... [and] accidents are continually occurring which maim the workers and serve to incapacitate them ... the unhealthiness of the occupation is beyond question; the low lights with which we are compelled to work, the confinement which impairs vitality, and the unseen dangers that menace.... [162]

The union urged the Alberta government to pass adequate compensation laws, to institute the eight-hour bank-to-bank regulation mirroring the one that already existed in the British Columbia mines, and also called for more frequent visits from the Mine Inspectors to check into the conditions in the mines.

The last resolution passed brought attention to the fact that the miners could not rely on receiving "dynamite or patent powders in a condition fit to use in their working places." This was an especially vexing problem for the miners, as it affected their time and earning power trying to bring down enough coal to make ends meet. The problem also cost lives. Several accounts throughout the Pass detailed the gruesome results of miners being "compelled under present conditions to thaw [the] powder." A coloured miner named George Washington had his arms and legs blown off at Coal Creek in 1902 when thawing dynamite over his stove; a year later an unidentified Slav suffered the same fate. At the time, the Crows Nest Pass Coal Company took the ineffectual step of stating that any employee found with dynamite in his home would be dismissed.[163] This announcement neither addressed the main problem nor stopped men from trying to bring the powder to a state where they could make use of it in the mines: Hiram Thrall was literally blown to pieces while warming the dynamite in 1905. The inquest verdict attached a rider urging the government to investigate the best ways of going about this dangerous task.[164] Since the union passed a resolution regarding this issue in 1908, the problem was obviously still extant.

The full spectrum of how miners dealt with dangers in the mine was demonstrated in their responses to bumps and blowouts. Despite their common occurrence, or perhaps because of the prevalence of these events, bumps and blowouts caused great consternation to both miners and mine management alike. Bumps and blowouts could regularly cause widespread panic among the miners. At one bump at Coal Creek in November 1917, the workers ran blindly out of the mine in a panic, without their dinner pails or coats. Mine management found the men outside the mine so scared that they found it almost impossible to get an intelligible account of what had happened.[165]

Blowouts in particular could cause major damage with the sudden release of methane gas: "... once an explosive wave gets started, anything is liable to happen in the Crowsnest Pass mines."[166] Correspondence flew back and forth between the mine inspectors, mine management, and government officials, all trying to come up with some way to either eliminate the bumps and blowouts, or at least to minimize the damage when they happened. The only method that appeared to work to some extent was to drill holes in the coalface where such a disturbance was likely to occur in order to relieve the pressure behind the wall: "... the whole question devolves on relieving the pressure ahead of the working face. If this cannot be done, it would be inhuman to ask or allow a miner to take chances such as are not experienced in any other industry."[167]

These safety measures, however, were dependent on each individual miner's attention to his working place, to his expertise in working the coalface, and to his inclination to report the first signs of a possible blowout. Gino Busato summed up many miner's reaction to bumps in the Coal Creek area: "... it used to bump like the dickens. Mind you when it bumped like that it was easy coal. You didn't have to dig it, you just had to shovel it."[168] Easy coal. What every miner wanted and what many were willing to achieve at the cost of waiving the danger inherent in a bump or blowout. While some miners could or would not stay in a mine that bumped continually, with attendant loss of life, many other miners chose instead to "court the blowout."

Bumps and blowouts rarely occurred without warning. Some miners reported that a periodic knocking occurred in the coal for days before a blowout; others reported that the coal would turn hard or tough, indicating a bump would happen in the next few days. When the coal turned hard, it was more difficult to work; on the other hand, if the miner waited for a blowout, it could produce tons of dislodged, easy-to-load coal. Although mine managers and mine inspectors tried to impress on miners the necessity of reporting these occurrences, to their exasperation, their warnings had mixed results: "There are plenty of indications if they were observed but these indications like many things more are left to 'everybody' or 'anybody' with the result that what is everybody's business is nobody's buisness [sic]."[169]

The mines of the Crowsnest Pass were dangerous. Accidents and fatalities could occur for a myriad of reasons: bumps, gas, coal falls, blocked chutes, haulage accidents, blackdamp, major and minor explosions. An analysis of miners' reactions to these dangers proves difficult since it is never black or white or total. Each miner responded in his own way although some patterns of habit and thought do emerge. Some men reacted by leaving the area; most stayed. Some felt hopelessness in their jobs that could lead to a greater chance of accidents; others enjoyed the work. Some responded to ambiguous messages on the part of management by taking more chances. Many followed all the rules and worked in a safe manner, yet died anyway in a chance encounter with a pocket of gas, slipping in the path of equipment, or buried under an erstwhile "safe" coalface. The tradition of sons following their fathers into the mines reinforced not only the passive response of a sense of fatalism, but also the active response of an awareness of pride and independence. In the worst-case scenario, these responses to danger could produce fatal results. In the best case, a sense of community was created, fostered by a shared awareness of camaraderie and brotherhood, independence and pride.

2 DEATH

When the damp explodes in a distant room
Or the roof of an entry falls,
Sealing the men in a living tomb
With thick and soundless walls,
When the women crowd at the open shaft
And wail as the women do,
It's then we call for the nerve and craft
Of the boys of the rescue crew!
They take the smoke as sort of a joke;
They dare the firedamp too;
For it's all their trade but they're not afraid,
The boys of the rescue crew![1]

— T.W. Davis, Carpenter, Builder and Undertaker
Repairs done Promptly, Estimates Free, Hearse for Hire[2]

On 24 February 1917 William Archer arrived for work as usual at the Greenhill Mine in Blairmore. A careful workman, well known and respected in the community, Archer had been a miner for many years and a member of the United Mine Workers of America since 1905. He was also one of the oldest inhabitants of Blairmore and had served in both the South African and First World Wars. On this day, however, Archer's luck in surviving dangerous occupations ran out. While working alone oiling cable rollers, a series of empty coal cars struck and severely injured him.[3] Some minutes passed before anyone noticed anything wrong, but then his fellow workers rushed to assist him. They first notified the pitboss and then the fireboss, William McVey, who hurried to meet the men coming out the mine carrying Archer on a stretcher. McVey put a blanket under Archer's head to make him comfortable and helped him to drink some water. The doctor arrived twenty minutes later, but Archer had already died of his injuries. In addition to several deep gashes around his head, the coal cars had also succeeded in nearly severing one foot.[4]

Mining accidents typical of Archer's inevitably resulted in a loss of life. Two professional and specialized responses to accidents and death evolved in the Crowsnest Pass during this period: mine rescue and the establishment of the undertaking business. Both of these responses were themselves influenced by larger events taking place in the western world. Miners, mine owners, and governments gradually recognized the need for better mine rescue techniques and equipment. The reality was that men could essentially do very little to alleviate the results of either minor or major accidents, but they tried. Mine rescue techniques, equipment, and training became increasingly professional. However, the effectiveness of this response could certainly be called into question.

The professionalization of working with the dead became the second specialized response. Several undertakers served the communities on both sides of the Crowsnest Pass during this period. While the professionalization of undertakers and the emergence of mortuary science proved a standard response throughout the western world at this time, the individuals in the Pass present an almost textbook case of this development. The constant necessity of dealing with mining deaths may have provided a catalyst for these men to pursue the professionalization of their trade. When major disasters resulted in mass death, the professional response of the undertakers became an important factor in how the community of the Crowsnest Pass dealt with the aftermath of tragedy.

William Archer's fatal accident illustrates a typical example of the rescue response by both miners and mine administration. The timeline of these events rarely varied. An accident occurred in the mine, whether a coalfall, a minor explosion, or a haulage mishap. The initial cries for help, from either the wounded man or his partner, alerted other miners and brought them running to the scene. As the first response team, these men usually secured a crude tourniquet, plied the injured man with alcohol or water or covered him with blankets. A call would go out for the doctor, either through an internal telephone system, or by sending a man running to the surface. Men transported the injured worker to the mine entrance on a stretcher or in an available coal car. The doctor usually met the transport party on their way out of the mine. Depending on his injuries, the wounded man would then either be taken to the wash house for the doctor to examine him or be moved immediately to the hospital. If, as often happened, the injured man had died during the trip out of the mine, his body would then be sent to the undertaking establishment to await the coroner's inquest and investigation.

While Archer's accident and subsequent death illustrates a typical sequence of events, it also provides a prime example of the problems associated with individual mine rescue response. The accident's location could prove problematic depending on the miner's relative isolation from the rest of the workforce. Archer worked alone, and only when the other men noticed the line of empty cars did they think to find out what had happened to him. Similarly, Steve Billesky suffered an accident in the McGillivray Creek mine while working alone in a left-hand crosscut while his partner worked the right-hand side. A coalfall buried Billesky, but in this instance, his cries for help immediately brought his partner running to his side.[5] A repairman named Johnson working alone in the same mine did not have the same luck. He died following a minor explosion when he took a wrong turn and ended up alone in a passage filled with gas. He never had a chance to call for help, even if someone had been nearby to hear him.[6]

Should an injured miner still manage to attract the attention of his workmates, an able-bodied man could still take an hour to reach the surface seeking help.[7] Mike Koziec received what would prove fatal injuries in the International Mine at 1:00 p.m., but it took two and half hours to transport him to the surface, "owing to the distance he had to be carried."[8]

The first men to respond to a distress call were frequently handicapped in giving first aid since they often found themselves in total darkness. A coalfall at the McGillivray mine extinguished the lamps of the two men working in the same area. Unable to see or locate his silent partner, the other man crawled out over the fall in the dark and ran for help. When he returned, his partner had already died.[9] In another accident at the same mine, a coalfall smothered the lamps of two miners. Although one man managed to relight his lamp and summon help, the delay no doubt worsened the shock that eventually killed his partner.[10] Restricted light could also hamper first aid attempts. At the inquest into the death of Albert Nycoma, the mine manager described the difficulties he encountered when trying to aid the wounded man: "The deceased was in the dark and the trip [mine cars] restricted the light from the surface."[11]

Even if an injured man could summon immediate assistance, there was no guarantee that the men who initially arrived at the accident scene possessed the knowledge necessary to apply the correct first aid. One contemporary writer lamented that:

> Numerous cases are on record of workmen who have met with serious accidents, such as the severing of limbs and compound fractures, in which arterial bleeding occurs, and who have lost their lives just because the person first on the scene lacked the knowledge of where to put a little pressure on the vital point, or who did not know how to adjust a tourniquet and so prevent the flow of the unfortunate victims' life blood.[12]

Miners died regularly from blood loss resulting from treatable injuries such as compound leg fractures.[13] Many accidents were far more severe. Tom Gibos' brother died in Coleman after losing an arm in an equipment accident. He "bled to death because the man with him didn't know what to do."[14]

Should someone actually know what to do in the event of an accident, lack of supplies could hinder proper treatment. As early as 1902, a writer reviewing coal mining in the Crowsnest Pass area commented on the general dearth of necessary supplies:

> The appliances required at every colliery or public work includes stretchers, boxes containing splints, bandages, tourniquets, needles, thread, antiseptics, oil paper, sticking plaster and other odds and ends. These requisites should always be kept in a place easy of access.[15]

The Alberta Department of Public Works took up the call three years later, stating: "All mining companies employing more than 20 men should have an ambulance room with stretchers and all appliances for first aid to the injured."[16] As late as 1928, however, when Mike Koziec broke his leg his partner had to "[improvise] splints from whatever material there was to hand."[17] Ambulance rooms were usually located outside the mine entrance. Many mine owners complied and instituted ambulance rooms, keeping supplies not only at the entrance but also at intervals inside the mine.[18] However, a continual supply could still never be guaranteed:

> Inspector Johnson in a statement urged that in the event of underground first aid stations being established in the mines, the men should co-operate in every way to see that the material left in them to meet emergencies should be taken care of and not stolen, as had been the experience in mines in districts in which he worked.[19]

Should a miner survive the accident and receive adequate first aid, the trip to the surface could still prove fatal. A falling timber fatally injured William Dunn in the Michel Mine in April 1918. Although Dunn lived some considerable time and reached the doctor alive, the inquest into his eventual death provides clear evidence that the trip out of the mine, down the hill, and into the doctor's office could have contributed to his demise. During an argument that erupted between the coroner, the mine inspector, the inquest supervisor, and the doctor regarding whether the time it took to bring an injured man out should become part of the investigation, Robert Strachan, the mine inspector, stated clearly: "I think certain men don't know a [deleted] thing how the men are brought down from the top of that hill; even a man with broken ribs; why it is hard enough to kill a horse." The doctor backed Strachan up by emphatically declaring:

> I would strongly suggest giving the matter attention. It is dangerous being carried down that hill, not only the hill but the way they are carried down; the movement and danger of a man slipping when walking. Then it takes so long for them to be brought to the hospital and they get too cold. [20]

If the miner survived the trip out of the mine, the wait for the doctor might still contribute to his death. The McGillivray Creek mine in Coleman shared a doctor with the Greenhill mine in Blairmore. In the case of accidents happening simultaneously at both mines, a considerable time elapsed before the doctor could attend to both injuries.[21] Breakdowns in communication also resulted in lengthy waits for assistance. After a coalfall injured John Bielec, his partner gave him "all possible first aid"[22] at the scene. Although immediately summoned, the doctor happened to be in town on a professional call. Upon receiving word of the accident he went to wait for Bielec at the hospital. The men carrying the injured man waited for a considerable time at the mine entrance before they learned they were to transport Bielec to the hospital.[23] In other cases, the doctor arrived at the mine entrance only to find no one in charge to give directions. When Dr. Tom Tolmie responded to a call for medical help:

> ... apparently no-one could direct him to the scene of the accident as he state[d] in his evidence that he waited at the head of the mine for word. Word was brought to him that the victim was dead and the doctor did not see the deceased until he was brought out of the mine.[24]

Despite the continual problems experienced during individual mine rescue attempts, programs to respond professionally to mine rescue needs were not instituted until massive mining deaths occurred elsewhere in the world. On 10 March 1906, the Courrieres mine in France exploded, sending 1,099 men to their deaths. Miraculously, thirteen men were rescued after twenty days' imprisonment underground and another was found alive five days later. Propelled by this disaster, the British Royal Commission on Mines stated in 1907,

"We are of the opinion that the question [for compulsory mine rescue equipment] is ripe for the development in this country, and demands the serious attention of the industry." Although the British government did not at this time decide to legislate the requirements for mine rescue equipment, by 1910 a number of mines had privately instituted mine rescue stations. Two years later, the Rescue and Aid Order of the *Mines Accidents Rescue and Aid Act* laid down the guidelines for the voluntary provision of appliances and training of men in the United Kingdom. [25]

Mine rescue operations took a similar path in the United States. A 1907 explosion in the Monongah, West Virginia mine killed 361 men. In reaction to this disaster, Congress allocated monies to create the federally run Bureau of Mines in 1910. The Bureau's mandate included scientific investigation into the research, development, and establishment of mine rescue equipment and stations.[26] Although the Bureau of Mines could and did make recommendations concerning equipment and training, the bureau could not force mine owners to follow their recommendations; individual states, not the federal government, had responsibility for regulating mines. Nevertheless, as in Great Britain, many of the American mine owners voluntarily instituted mine rescue stations, equipment, and training at their operations.[27]

Although Canadian mines also had their share of major disasters that might have served as catalysts, the institution of mine rescue stations in Canada evolved more slowly than in either Britain or the United States. The 1902 explosion in the British Columbia Coal Creek mine resulted in 128 dead, while in the Nova Scotia mines, at least 124 had died in a series of explosions before 1900. However, in Canada as in the United States, regulation of mines did not fall under the federal mandate. By 1909, only British Columbia had legislation in place requiring coal operators to keep mine rescue apparatus at their mines. Despite this, coal mine owners in British Columbia and in other provinces frequently instituted training on a voluntary basis. They were motivated not only by a concern for the protection of their employees, but more often by their own self-interest. While prompt and efficient attention to accidents lessened the injuries to the men, it also meant "the loss of the injured workman's time [was] reduced to a minimum,"[28] thus ensuring an uninterrupted production schedule.

In British Columbia, the *Coal Mines Regulation Act* of 1911 enforced the presence of some form of mine rescue equipment at all times:

> There shall be established by the owner, agent or manager of every colliery such number of oxygen helmets or some form of mine-rescue apparatus as may be approved by the Minister of Mines. Such mine-rescue apparatus shall be constantly maintained in an efficient and workable condition, and shall in all cases be so stored or placed in or about the mine as to always be available for immediate use.[29]

A permanent mine rescue station was built in 1912 at Fernie. By the end of that year, the four principal British Columbia coal mines (Hosmer Mines Ltd., the Crows Nest Pass Coal Co. Ltd. at Michel and Coal Creek and the Corbin Coal and Coke Co. Ltd.) could list the following equipment on hand for mine rescue work: a total of twelve sets of two-hour Draeger apparatus (breathing machines to assist in gassy mine rescue work); three oxygen pumps; twenty-four oxygen tanks; and four pulmotors.[30] The pulmotors were devices used to restore breathing to men who had succumbed to gas or smoke inhalation. Although none of the mines had mine rescue training stations, regular voluntary training did take place inside the mines. This list of equipment, however, indicates that the rescue equipment available was totally inadequate for the needs of the over seven thousand men employed in the mines at this time.[31]

Alberta mines did not have any statutory regulations for compulsory mine rescue equipment or training during this period. The first *Mines Act* of 1906 included only a single brief section on ambulances: "Properly constructed ambulances or stretchers with splints and bandages shall be kept at the mine where more than twenty men are employed ready for immediate use in case of accident." These instructions appeared under the "General Rules" of the Act and were only required "to be observed so far as is reasonably practicable in every mine."[32] A revision of the *Mines Act* in 1913 did at least remove the ambulance clause from the "General Rules" and re-institute it as a "Section" in its own right. However, apart from repeating the ambulance clause, the legislation changed very little.[33] Collaborative efforts between the province and the mine operators did manage to result in the establishment of the first mine rescue station at Blairmore in March 1912. The provincial government and the coal mine operators on the Alberta side of the Crowsnest Pass shared equally in the costs for this three-room structure.[34] One room contained storage for equipment, the second was an office while the third contained a specialized "smoke-room" for training purposes.

Unlike individual mine rescue operations where one or two men were hurt and required urgent assistance, the need for adequate equipment and trained men rose exponentially in the case of a major mine explosion. The confusion that reigned both inside and outside a mine following a massive explosion became an important factor in driving the need for organized mine rescue. A report from an experienced miner during a 1916 explosion at Coal Creek graphically describes the confusion that existed inside the mine:

> I was struck on the back by a violent wind, dust and debris blast that knocked me down violently against the rib side of the roadway. At the same time my light was extinguished.... The last movement was so violent that it blew down doors in the region where I was stuck, reversed the intake air currents and flooded the main tunnel with gas and dust clean out to the pit mouth. How I passed through this surcharged mass of gas, dust, [and] dangerous mixed air that stagnated the turmoil for the 4,000 feet I traveled through God only knows, for my mouth, ears, eyes and nose were all packed solid when I reached the surface.[35]

Multiply this one man's description a hundredfold with men stumbling around in the dark, struggling to reach the surface, either injured or dying, fighting gas, smoke, and debris, and the need for organized mine rescue techniques and equipment becomes obvious.

The 1910 Bellevue explosion that killed thirty-one men provided several prime examples of the problems that could frequently occur in mine rescue operations during a major disaster. Immediately following the explosion, a call went out for equipment and trained men to be rushed in from the established mine rescue station located at Hosmer, British Columbia. The men, equipment, and supplies arrived at Bellevue by train within four hours of the initial call, having stopped at Coal Creek to pick up additional supplies and equipment.[36]

However, on arrival, the rescue crew found massive disorder at the mine mouth:

> There did not seem to be any officials of the company meeting us ... we decided to get one of the mine officials but we could not get anybody, neither could we get a plan of the mine as it was locked up in the superintendents desk.[37]

Eventually finding their way into the mine, the rescuers discovered that the same confusion reigning outside was also evident within. They found their initial passage impeded by "quite a few people who appeared to be merely sightseers ... there were quite a few men but there was no attempt at discipline."[38] The Hosmer team doggedly pursued their own mandate and eventually established a base of operations inside the mine from which to send out further rescue parties. Additional problems arose when some members of the rescue party went too far and too fast, placing risky pressure on the limits of the rescue apparatus. One of the rescuers found himself in the unenviable position of having to rescue an earlier party:

> I immediately went inside to where the men were and found them all unconscious and down on the floor except Alderson who was just able to speak. I immediately realized the situation and returned back to where the rest of the party was and told them of the condition the men were in and that it was impossible for me to put the ½ hour apparatus on the unconscious men and it was impossible for me to carry them as the oxygen in my machine was now only good for a ½ hour. So then somebody decided they would make an attempt to rescue the men by placing ourselves 10 feet apart and thus bring the men outside [to] fresh air.[39]

A final statement by another of the rescuers painted a damning picture of the lack of organization, communication, equipment problems, and training procedures evident during the Bellevue rescue attempt:

> The accident to the rescue party was caused by them going ahead of the ventilating current in order to try to rescue the men that the Draeger men had located alive at the face. I don't think the Draeger apparatus was used to advantage and it was the indirect cause of all the trouble to the rescue party. I consider that the rescue work suffered greatly owing to all the chief mine officials being in the

exploring party and there being no-one with sufficient authority left behind to issue directions from the base of operations and to control the work in the parts of the mine secured by the pioneer rescue party.[40]

The problems at Bellevue were not an isolated case. Nor were these the only problems associated with a mine rescue during a major disaster. Massive explosions frequently resulted in extensive damage to the mine workings. When a portion of Turtle Mountain slid into the valley in 1903, the resulting debris completely covered all entrances to the Frank mine. The men remaining inside the mine ran to all the known exits, only to find them completely blocked. The main entrance had completely disappeared; they found the airshaft choked with debris and the lower tunnels rapidly filling with water.[41] The trapped men eventually dug themselves out, emerging at a point beyond the area where the rescue workers were frantically digging with their bare hands, picks, and shovels.

The Hillcrest mine explosion of 1914 also caused major damage to the mine entrance. Mine rescue attempts were initially hampered by the fact that one entrance to the mine had been completely destroyed.[42] When the rescuers did manage to clear away the debris that blocked the entrance, a far greater amount of devastation inside the mine spread before them:

> With the greatest difficulty the gangs of rescuers ... can make any headway. Men, horses, timber, rails and wagons are jumbled in chaotic mass and the path is so strewn at every step with the debris that only those men who were fortunate enough to have been working near the pit mouth have any chance of being brought out alive.[43]

Rescue difficulties could also be compounded by environmental factors. The 1916 explosion at Coal Creek occurred during a terrific thunderstorm. The heavy wind and rain, jagged bolts of lightning, and the deafening effects of the thunder hampered the efforts of the men trying to dig through the debris. The storm knocked out communication lines and prevented word of the disaster reaching Fernie.[44] An artificial environment caused by the force of the explosion could also affect rescue efforts. The energy of the wind caused by a 1909 explosion in the Morrissey No. 1 mine prevented any rescue attempts for a precious thirty-five minutes.[45]

The confusion and resulting panic of any major explosion created severe organizational problems. Men on the surface did not wait for a rational, organized response but instead acted on impulse. Following the explosion that would kill seven men at Michel in 1904, Superintendent Wilson and other willing men "forced their way into the deadly afterdamp with the hope of rescuing their comrades."[46] Likewise, William Hutchinson, the mining engineer at Hillcrest, immediately braved the mine after the 1914 explosion in an attempt to rescue survivors.[47] Owing to their zeal in entering the mine before the deadly gases had dissipated, both Wilson and Hutchinson almost became victims themselves, eventually staggering out into the open air, only to try again. Other men, finding themselves safe from harm, nevertheless succumbed to the dreadful emotions caused by finding friends or family missing. Like Henry Grewcutt in the earlier explosion of 1907, David Murray was also a father who escaped the initial explosion at Hillcrest in 1914. However, upon learning that all three of his sons were still missing, Murray eluded concerted efforts to restrain him and plunged back into the mine to search for his family. All four Murrays were later found dead.[48]

The fact that accounts of men running into the mine to assist those trapped are repeated over and over again is an obvious tribute to the men themselves. These men were volunteers who immediately, and often without thought, offered their lives and their strength to aid their fellow miners trapped in the mine. After the 1902 Coal Creek explosion, the call for volunteers brought men from Fernie and the surrounding areas in an effort to reach those trapped inside while there was still chance of life.[49] These actions were repeated during the Hillcrest explosion of 1914:

> When it seemed that the mine was doomed a call came for more fire-fighters. In spite of the possible danger of another explosion from flames skipping along the ribbons of methane gas and encountering a pocket of gas, more men went into the mine. They were too tired to be heroes. They just wanted to finish their gloomy task.[50]

Despite the usual glowing reports from newspapers, eyewitnesses, and historians alike, there are a few jarring notes. Although these accounts are primarily of selfless action and self-sacrifice on the part of miners and management, there are other stories. Following the 1902 Coal Creek explosion, several reports make note of the fact that the Slavs were noticeably absent from any of the rescue parties. This is rather remarkable, since the majority of the workers inside the mine who died that day were members of the Slavic community.[51] In another instance, Duncan McDonald, the superintendent of the Mine Rescue Car dispatched to assist in the Hillcrest disaster in 1914, wrote:

> I made a call for to [sic] muster a volunteer mine rescue team just to be held in readiness on the outside. But I did not get a single response from a crowd of over 200. A Lethbridge team had to fall in even although [sic] they were completely exhausted with their continuous work.[52]

Although these instances are rare, and are far outweighed by the unselfish actions exhibited in countless other rescue scenarios, the occurrences are worth noting.

Sometimes the force and extent of an explosion resulted in the death or incapacitation of those who were supposedly in charge of organizing the mine rescue. The explosion at Hillcrest decimated the upper echelon of command. The blast almost succeeded in completely negating any prior knowledge or preparedness that these men may have possessed:

> Engineer Hutchinson was the only one with sufficient knowledge of the mine workings to give any clear picture of what had happened. Fireboss John Ironmonger was still unconscious; fireboss Sam Charlton was unaccounted for; Mine Superintendent James S. Quigley, pit boss Thomas Taylor, and other men who had intimate knowledge of the mine were still missing.[53]

A bad situation immediately became worse when Hutchinson impulsively dashed into the mine immediately following the explosion. He was completely unaware that all the other men with the authority to direct rescue attempts were incapable of action or missing. Only when Hutchinson stumbled out again into the fresh air did he find "a mine rescue crew anxious to go inside, and wanting a plan of the mine workings. I got them a blueprint from the mine-managers office, and arranged to have more made and sent."[54]

All of these factors that could and did go wrong during the aftermath of a major explosion had already been recognized and analyzed. The Government of Canada's Commission of Conservation mandated W.J. Dick in 1912 to write the first comprehensive report on mine rescue work, equipment, organization, and training.[55] Dick produced a thorough overview of the history of mine rescue and succinctly reported on the current state (or lack of) mine rescue work in Canada. He went to considerable lengths in the appendices to illustrate how a mine rescue operation ought to work. Dick had a clear understanding of the urgency that a mine explosion called for, noting: "experience has repeatedly shewn that the effects of a serious disaster may be minimized, both as regards life and property, if *immediate assistance* can be given" (Dick's emphasis).[56] Dick identified extensive training as one of the top priorities in running a successful mine rescue operation.

Dick's suggestions for training were soon adopted at the new Mine Rescue Stations located at Blairmore and Fernie. The training schedule consisted of six days of intensive work, starting with hands-on familiarization with equipment. The men learned both how the various appliances operated and also how to take them apart and repair them. The trainees then spent long periods of time wearing the equipment in order to get used to the weight of the Draeger devices. For intervals of anywhere from one to three hours, the men walked and climbed over obstacles to simulate the conditions they would find in a mine after a major explosion. They also performed basic mining repair skills such as sawing props, erecting brattice screens, and moving rock, all while wearing the cumbersome equipment in a smoke-filled room.[57]

Dick's recommendations also included specifications concerning the types of men who should be encouraged to join the mine rescue crews: "They are to be sound in limb, with perfect hearing, of non-excitable disposition, duly certified as qualified ambulance men and thoroughly acquainted with the underground workings of their respective pits."[58] His apprehensions regarding the mental stability and disposition of the rescue crews were understandable. Mine rescues required unpleasant work in deadly surroundings while operating under severe time constraints. The faint-hearted or excitable would only prove a danger and a burden to the rest of the team.

Kenny Rumley, an experienced miner from Nova Scotia, joined his mine's rescue team in 1914. He remarked:

> A man can be as brave as anything on the surface. He can learn in school how to do something, but when it comes to doing it, that's a different thing. There are good men, able men, but you don't know what's going to happen when they get into danger.[59]

During an early rescue operation, one of Rumley's men "went hysterical, trying to tear off his helmet." The others managed to wrestle him to the ground and subdue him. In instances where time was precious and the men depended on each other's actions, clear and cool heads were an absolute necessity. Duncan McDonald, the superintendent of the Lethbridge Rescue Station, which assisted at the Hillcrest explosion in 1914, also commented on the unsuitability of some of the men assigned to his rescue crew. One he cited simply as "foolhardy" and "incompetent" for taking off the apparatus in a poisonous atmosphere and risking his and other's lives. McDonald eventually took another man off active rescue duty, pronouncing him "very excitable and ... not able to control himself."[60]

Dick identified physical ability as a second vital requirement for rescue workers. The Draeger devices the men carried into the mines were heavy and cumbersome (see photo 2). Dick's concerns were deep enough that he considered it necessary to include a "special warning" in his report. He stated bluntly that if a man had not trained properly or was unused to the necessity of breathing and moving slowly, the device could easily become a "DEATH TRAP" (Dick's emphasis).[61] The breathing apparatus weighed about thirty pounds and contained a reservoir of pure oxygen. When a rescuer first put on the device:

> The wearer takes a deep breath. The nitrogen in that breath he breathes over and over again for 2 hours. As the air is exhaled from the lungs it passes through a potash filter which absorbs the carbonic acid gas and leaves the pure nitrogen as a vehicle for a sufficient supply of oxygen escaping from the reservoir to fill the lungs again.[62]

2 Draegermen with Equipment Rushed to Scene of Bellevue, 1910.
Provincial Archives of Alberta, A2039.

On some rescue missions, men carried two of the devices in order to bring breathable air to miners trapped in the dark. The intense pressure and physical cost of wearing a device on his back, navigating over obstacles and through smoke-filled tunnels while carrying another thirty pounds of equipment, could prove too much for some men; Fred Alderson died during a rescue attempt following the 1910 Bellevue explosion:

> Alderson started in in his outfit and carrying another full suit for Mr. Strachan's use; the load was, however, too much for him and he dropped it on the way ... this appears to show that Alderson must have hurried in, and made a greater demand on the apparatus for air than it was capable of supplying.[63]

District Mine Inspector Evan Evans died in a similar manner following a tragic sequence of errors after the 1915 Coal Creek explosion. After entering the mine to inspect the damage, Evans apparently became convinced that the party could not accomplish anything and called to the others to retreat. When Overman Robert Adamson heard Evans shouting, he assumed Evans was in difficulty. Adamson removed his own mouthpiece, against all training and common sense, to communicate to the men ahead of him in the mine. The gas overcame Adamson and when Evans reached him, he attempted to carry Adamson to safety. Evans soon collapsed himself under this burden and both Evans and Adamson had to be dragged into the fresh air by the returning inspection party. Although Adamson responded to the efforts of the pulmotor, Evans did not. He died despite the efforts of the doctor who spent four hours trying to revive him.[64]

Upon hearing of the circumstances of Evans' death one observer commented caustically: "Never let a big stout man don a rescue apparatus. If he goes down, a pony is needed to drag him out or the other fellows are lost trying to save the one."[65] Although this was perhaps an incautious remark, the facts in the case were underlined by the official report of the fatality to the British Columbia Minister of Mines in 1915. The otherwise dispassionate report makes an uncommon point of commenting on the size and weight of both Adamson and Evans. The report goes further to state that following Evans' autopsy, although his heart was found in fair condition "there was considerable fat around the heart, which would materially affect its capacity for work."[66] These comments and observations should not detract from Evans' considerable knowledge and experience of the mines. However, the fact that the comments were made underscores Dick's recommendations regarding the types of men that were the best suited for mine rescue work.

Even if the rescue crews were composed of hardy competent men, organization and communication remained a vital component in ensuring a successful mine rescue. Dick strongly recommended that the guidelines stipulated at the National Mine-Rescue Conference in September 1912 be followed closely to ensure adequate organization.[67] These guidelines stated that at least two sets of teams were needed, one for "inside" work and one for "outside" work. The outside organization would ensure that only those with business inside the mine would be permitted to enter, all gear inspected, supplies and provisioning areas established, and a safety cordon drawn around all damaged areas. The inside organization would first send an advance group into the mine to determine the damage and the probable locations of any survivors. This team would then set up a base of operations in a safe area from which they could send out other rescue parties, all the while maintaining constant communication. The final stipulation for each group insisted: "Strict discipline should be maintained at all times."[68]

On paper, Dick's recommendations for mine rescue operations appeared admirable. In reality, the complete chaos and shock of a major mine explosion often negated the finer points of his organizational model. The force of the blast could kill most of those caught in the immediate area, while the blackdamp and afterdamp could soon catch up with those who might have initially survived. Mine explosions were extremely powerful. One of the Morrissey mine explosions in 1904 displaced an estimated "3,500 tons of coal dust in the face of a current of 57,000 cubic feet of air per minute," while a 1909 explosion deep in the mine still managed to extinguish a safety lamp located three hundred feet outside the main entry.[69] The 1914 Hillcrest mine explosion "hurled Charles Ironmonger ... 60 feet through the air ... the roof blew 40 feet backwards, and the 8 inch concrete wall shattered on impact."[70] For men inside the mines during these massive explosions, primary injuries to the internal organs caused by the blast wave or pressure to the body were usually enough to kill them. Secondary injuries resulted from blunt or penetrating trauma from flying debris or from the blast wave slamming the miners into solid objects.[71]

When viewed analytically, the actions of the men who did not wait for formal organization before rushing off in search of the injured or trapped appear foolhardy and often suicidal. However, their actions, often taken without thought for their own safety, were frequently more successful in rescuing survivors than the actions of those who had waited for orders and supplies. After the McGillivray mine explosion in 1926, the pit boss immediately "led a party of eleven down into the mine with no other protection than a canary."[72] They were successful in finding several miners and leading them to safety before the rescue party even arrived. As a survivor of the 1914 Hillcrest explosion, Yuriy Frolak owed his life to the mine rescue crews. Although he and his friends had stumbled and crawled as far as they could on their own, "the reserves which had driven us to superhuman efforts were finally exhausted. We collapsed and sank into oblivion to be eventually brought out by rescue teams."[73]

Few men were able to act rationally in the precious few minutes after a major explosion when organizational skills were needed the most. The rare examples of those who could maintain their composure soon found their way into local celebrity. True Weatherby[74] was one such man. Immediately after the Coal Creek mine explosion in 1902, Weatherby, whose previous claims to fame included a barroom brawl and a ten dollar fine, kept his head: "In the first moments of partial, if not complete, paralysis of the official leaders, his brain and hands were kept intelligently busy."[75] The first to enter the mine looking for survivors, Weatherby also volunteered immediately for the initial relief party organized by the superintendents. He became "a hero in the company of heroes. There appeared to be no limit to his endurance. For twenty-three hours he held to the work of rescue with marvelous tenacity performing the work of two ordinary men."[76] At Hillcrest in 1914, reversing the ventilation proved vitally necessary to save the men trapped and trying to reach the surface. The mine manager, John Brown, earned praise for his instant ability to realize the necessity for this action: "Moments after the explosion Mine

Manager John Brown reversed the fans so that the gas laden air would be pulled out. It was this quick decision that helped to save some of the miners desperately trying to make their way out to safety."[77]

The problems associated with mine rescue operations in the face of overwhelming disaster may be best explained by fitting the events into the following model for time phases of a disaster:[78]

Time Phase	Proximity of Danger	Coping Mechanism
Steady State	Distant	Preparedness
Crisis	Approaching	Crisis Management
Disaster Impact	Imminent/present	Survival/Rescue
Afterperiod	Passed	Working through Shock

Viewed from the standpoint of this model, complete preparedness for a mine explosion would have been a rarity. Preparedness, in the case of stockpiling supplies and equipment and seeing that men were trained, seemed the easiest role of the mine organization. Major explosions, while rare and always unpredictable, were not completely unexpected.

However, miners responded slowly to the government's attempts to interest them in taking courses in mine rescue techniques or first aid. In 1912, the Blairmore Rescue Station boasted that sixty men had undergone the complete course of instruction. However, the Alberta Mines *Annual Report* for that year still lamented "a considerably larger number have put in a few days and then discontinued for various reasons. At times considerable difficulty has been experienced in inducing men to undergo the course of training."[79] Two years later, only 142 men could be listed as holding mine rescue certificates in all the mines in Alberta.[80] The total number of men employed in mines at the time amounted to more than 8,000.[81] Similar problems are evident from British Columbia; although the mine inspectors gamely make note of the number of men who did take the training, the British Columbia Mines *Annual Reports* also contained a continual litany of complaints that many others were not interested in learning mine rescue techniques.[82]

One of the initial problems in keeping men involved in the training regimen could have been that they were at first not reimbursed for the time they spent training.[83] A complete course of training over six days and the required return every month for a refresher course meant a significant loss of income to most miners. This was somewhat ameliorated in 1916 when the Government of Alberta agreed to pay "fifty cents an hour for every man required to undergo eight training's of three hours each." The Alberta Mines *Annual Report* for 1916 could report glowingly that "over five hundred men have been fully trained in the use of the apparatus and this number is continually being added to."[84] By 1921, these numbers had risen to include 1,388 men trained in mine rescue work and 980 in first aid.[85] In British Columbia, the government and coal companies took a little longer to realize that the men needed compensation for the time it took to train; by 1922, however, although the men were still expected to train on their own time, the companies would pay for the time and the work involved.[86]

More success could be seen in the numbers of men interested in first aid training as opposed to the more arduous and time-consuming mine rescue courses. The doctors employed in the mines initially gave first aid training on an ad hoc basis, but the lack of standards or a set training regimen hampered their attempts. Training was given a boost when the St. John's Ambulance organization became involved in the process around 1910. This group set standards, established regular classes, and sponsored evenings of "delightful and instructive entertainment" to entice the miners and the larger community to participate in first aid practices. The organization later employed the newly minted device of moving pictures to instill interest among the miners, as four hundred adults crowded into the Fernie auditorium to see films on "The Care of an Injured Miner" and "The Story of the Rock-Dusted Mine." St. John's also assisted in establishing intercommunity competitions throughout the Crowsnest Pass with prizes for the teams with the winning times or techniques.[87]

Another reason for the slow growth of interest in preparedness for mine rescue work could have been the pragmatic attitude of the miners. A reviewer of Dick's initial report in 1912 wrote of his hope that the report would "go far to remove from many minds the lurking suspicion which still exists that the provision of mine-rescue apparatus is a 'fad' advocated by enthusiastic but unpractical [sic] persons."[88] However, as most miners knew from experience, when a mine blew, time was the absolute crucial element. The odds of finding men alive after the first few minutes were extremely low. By the time a man had managed to put on the Draeger device and found his way to an injured companion, he would be faced with having to carry the weight of both an incapacitated miner and the equipment and still make it to the surface in the time allowed by the oxygen capacity of the device.[89] While the half-hour Draeger device was admittedly lighter, it severely limited the travel time allowed before a rescuer was forced to turn back.

Miners and managers both chose through inaction and attitude to ignore or downplay preparedness, the necessary coping mechanism for Phase One of a disaster. This lack of preparedness would only be compounded by the fact that Phase Two in the disaster model — crisis management before a disaster struck — did not exist in the case of mine explosions. Major explosions could not be predicted either as to their location, source, destructive power, or numbers of men involved. Miners and managers were therefore catapulted from the day-to-day routine directly into Phase Three of a disaster, where most men operated in survival mode.

Mine explosions instantaneously created massive stress problems for both miners and management alike. In the immediate aftermath of an explosion, men found themselves faced with immense physical and mental obstacles. These included their own severely reduced control over the situation, high uncertainty as to their own safety and that of others, an extremely low level or complete lack of available information, and a time pressure, second to none, for them to act effectively. In the face of these overwhelming stresses, the designated problem solvers, even if they were still able to physically function, found themselves faced with critical questions. The decision-makers had to immediately recognize the problem, gather all relevant information despite the extreme unpleasantness of its nature, decide on a path of action, and implement this action by directing others similarly affected.[90]

The effectiveness of mine rescue work during this period cannot be evaluated by counting the number of men saved, especially in relation to major mine disasters. The likelihood that any additional survivors could have been rescued from the major mine disasters (Coal Creek in 1902, Bellevue in 1910, Hillcrest in 1914, Coal Creek again in 1917, or Coleman in 1926) after the first few minutes appears very remote. The level of confusion at the mine mouth might be used to measure effectiveness of response. However, given the universal problems and stresses associated with mine disasters it seems obvious that the confusion of 1910 was still evident in 1914, 1917, and also in 1926.

Instead, an evaluation of the effectiveness of the mine rescue response may be possible by looking at the main components W.J. Dick pointed out in his initial report: training, organization, communication, personnel, and equipment. The increasing numbers of men who took part in mine rescue and first aid training would seem encouraging. Yet despite a boasted 25 per cent increase in membership at first aid competitions in 1922,[91] an article six years later listed four minor accidents that had proved fatal and underlined the lack of basic first aid knowledge.[92] As late as 1928, inquest juries were still calling for adequate stretcher or coal car components to bring men out of the mine faster following an accident.[93] At the same time, mine owners were evidently still reluctant to provide supply stations for first aid deep inside the mines.[94]

Although scientific advances and research resulted in the development of better breathing devices, the devices were of little use if operated improperly or if they were unassembled when needed. The Bellevue verdict issued in 1911 deliberately cited the need for more safety precautions related to using the Draeger devices.[95] This call for more vigilance appears to have fallen on deaf ears, as three years later considerable confusion reigned when the Draeger apparatus arrived at the Hillcrest disaster in a state of complete disassembly.[96] And while good communication was also a basic requirement for a successful mine rescue, inquest juries in 1914 still saw the need to call for better communication between the interior of the mine and the hospital.[97] Ten years later little had changed: an almost identical rider was attached to the inquest of John Bielic's death in 1924 "that in future medical aid be given at the mine as soon after the accident as possible."[98]

This is not to say that miners did not admire the bravery involved in being part of a mine rescue crew. Mine rescue was an extremely risky business and fully recognized as such. When all instincts were to leave the mine in case of further coal falls or explosions, mine rescue teams deliberately chose to enter the danger areas and put their own lives at stake. As one present-day mine rescuer remarked: "It took quite a bit to prepare yourself mentally to go down into that place and, once you were set to go, you had a hard time calming down again."[99] A commissioner delegated the task of investigating a 1926 explosion made deliberate note of the bravery of the mine rescue teams: "It seems fitting that some public recognition should be given to the really heroic conduct of all concerned from the lowest to the highest degree."[100] However, miners were more likely to rely on themselves, their partners and on the spirit of camaraderie and interdependence fostered in the mines to assist them in times of either individual accidents or major disasters. Watching each other's back became second nature, and when danger turned to real or potential death, most miners found it a comfort to know that help was always near at hand.

While some men were indeed lucky enough to be assisted out of the mine by the rescue teams, the Draeger-men were far more likely to find themselves involved in "recovery" work than "rescue." Recovery work included the clearing and repair of twisted and blasted tunnels, chalking off blocked passages where men might be lying, finding and transporting the bodies to the surface, and putting out any fires that might still be burning underground. The force of explosions caused such extensive damage that "recovery" presented a far more realistic view of the work than "rescue."

Miners knew that the chances of their rescue in a major disaster diminished with every minute that passed after the initial explosion. They also knew that miracles did occur. In November 1909, an untended fire raged out of control in the Cherry Mine in Illinois, resulting in 259 dead. Twenty miners survived the horrendous conditions underground and were brought out alive after spending eight days trapped in the superheated mine.[101] If they were trapped in the dark with no way out, and if immediate assistance was unavailable to deliver them to the open air, all miners still knew that eventually, the "dauntless men"[102] of the rescue crews would come looking for them. Despite the confusion at Bellevue in 1910 and the death of one of their own, the mine rescuers still managed to bring a number of men out alive.[103] At Hillcrest in 1914, William Hutchinson succeeded in finding six men alive, almost two and half hours after the explosion.[104]

Undertakers

While the work of the mine rescue teams ended after the bodies were brought from the mine, the work of other "recovery" professionals had just begun. The undertakers of the Crowsnest Pass, although they could not provide the hope of a miracle such as the mine rescue crews, were nonetheless instrumental in providing some comfort to the survivors. The demeanour, competency, and skill of the undertakers who served the Pass provide an almost textbook case of how the care of the dead became a professional business during the early part of the twentieth century.

Undertaking first emerged as a separate trade in England around 1688 when William Russell "undertook" to provide all the funeral services and accoutrements for Baron Fitzwilliam's son.[105] Previously, all arrangements for upper class funerals were channelled through the College of Arms. First established in the sixteenth century, the College had evolved into the institution solely responsible for the right to arrange and marshal the funerals of the elite. However, when the Crown refused to renew the College's warrant after 1688, undertakers began appropriating the business of death. They also sought to expand the trade by offering to arrange funerals for all citizens, whether upper, middle, or lower class.[106]

Since the commerce in death had unsavoury roots at best, more than two centuries would pass before undertakers gained any recognition and respectability as a profession. Undertakers had long been linked in the public's mind with the trade in corpses or "body snatching" for medical study. However, in 1832 the British government passed the *Anatomy Act*, which gave legal permission for medical schools to purchase unclaimed corpses, reducing somewhat the stigma attached to the undertaking trade.[107] Despite this, undertakers themselves were still rather derisively described as "bedmakers to the dead." One of the major problems undertakers faced was that they only took charge of the body when the "legitimate" professions of doctors and lawyers were finished with their own work.[108] Undertakers would have a long fight ahead of them before they were recognized in the same professional sense as these more established occupations.

Undertakers soon recognized a business advantage in the Victorian mania for pomp, ceremony, and the performance value of funerals. They began to market their skills to members of the emergent middle class who were eager to show off their new wealth and status through elaborate funerals. Embalming assisted the struggling undertaking profession considerably when the science of injecting preservation fluid into the body's arteries became widespread in the late nineteenth century. This new art of preserving a lifelike appearance allowed for a longer mourning period free from the imminent corruption of the body. More importantly, embalming provided undertakers with a significant marketable skill. Embalming rapidly established its own merits when it proved sufficiently akin to the talents required of recognized medical professionals. The new art gave the undertakers a legitimate science to call their own.[109]

Historically, most undertakers were unlikely to be solely in the trade of arranging funerals. Many undertakers started out in the building, carpentry, or upholstery trades, since these occupations were directly linked to the needs of catering to the dead. The new emphasis placed on the coffin as the primary element in a funeral greatly aided the undertaking profession as well. By the end of the nineteenth century, undertakers had added not only the supply of funeral "furniture" to their primary mandate concerning care of the body, but also the supplying of all manner of services required by the bereaved. Undertakers supervised and managed every aspect of bereavement from the time of actual death to the placing of the body in the ground.[110]

The backgrounds of undertakers in the Crowsnest Pass were no exception; all of them initially had at least one other primary trade besides that of undertaker. In Pincher Creek, the nearest large town on the eastern side of the Crowsnest Pass area, the Scott brothers were perhaps the most ambitious. They first advertised themselves in 1904 as: "Architects and Builders, Blacksmiths, Wheelwrights and Undertakers." T.W. Davies of Coleman, the longest-serving undertaker on either side of the Crowsnest Pass from 1909 to 1921, first promoted himself as "Builder and Undertaker," placing the emphasis in his early advertisements on his contracting and carpentry business. R. Elliott from Kaslo placed some early ads in the Fernie newspapers for "Furniture and Undertaking," but he was quickly supplanted by the partnership of Scott & Ross, whose primary business was also a furniture outlet located in Fernie. The eventual successor to first the Fernie and then the Coleman undertaking businesses was A.E. Ferguson, who had lived quietly in Fernie as a plumber before taking on his new profession in the 1920s.[111]

Since undertaking was rarely the sole profession for any of its practitioners, a bidding war for undertaking services in the Crowsnest Pass newspapers was not surprising. For a short time in 1908, Fernie witnessed a plethora of undertakers duelling it out with increasingly elaborate advertisements. While R. Addison's quietly worded and graphic-free ads eventually sank from view, William Scott went head-to-head with J.H. Reid. Scott advertised with an elaborately decorated hearse while Reid trumpeted that he not only had all the "most modern and up-to-date goods" but that he was also "in a position to handle funerals in the Eastern Style." Reid was apparently savvy enough to noticeably cater to the large eastern European and Slavic population in the Fernie and Coal Creek area.[112]

R. Addison was the first gentleman of the trade to advertise on the Alberta side of the border. He proved a bit of an early anomaly. Obviously a forward-thinking man, Addison advertised his services in both the Coleman and Fernie newspapers in 1908, not as an undertaker but more impressively as a "Funeral Director and Embalmer." The firm of Thomson & Morrison in Fernie also advertised themselves as "Funeral Directors" early in 1913.[113] The term "funeral director" is reflective of the larger and more professional role that undertakers were moving into. Funeral direction involved myriad services and not just the care of the body. The term "funeral director" was extant in the United States by 1884, and some of the Pass practitioners certainly appeared to prefer the term, to judge from its use in their advertising. However, the residents of the Crowsnest

Pass seemed to be more comfortable with the term "undertaker" as evidenced by T.W. Davies' continuing use of the term from 1909 until 1921. Even as late as 1928, a short editorial in the *Coleman Journal* stated:

> If you asked a man what his occupation was, and he told you he was a mortician, you might be excused for wondering what that might be. But it is just an up-to-date word for funeral director, more commonly known as the undertaker.[114]

In 1909, T.W. Davies took over the business from Addison as the principal undertaker serving the Alberta side of the Crowsnest Pass. His long service to the community exemplifies the growth of the industry in the early quarter of the twentieth century. Davies' introductory advertisement to the residents announced grandly:

> T.W. Davies Carpenter and Builder of Coleman wishes to thank his many friends for their kind patronage in the past and also wishes to inform the residents of Coleman and Blairmore that he has been induced to put in a stock of caskets and will in future be prepared to undertake all arrangements for funerals.[115]

Davies never completely abandoned his trade as a building contractor. Throughout the time that he was an undertaker, Davies continued to place bids and receive contracts for projects such as the new schoolhouse in Coleman.[116] One reason for this could have been that the undertaking trade, while steady, might not have been quite enough to support full-time living expenses. The difficulties of receiving remuneration were a constant problem. Compensation claims indicate that in a number of instances Davies had to wait a year or more for payment for his services. After Felix Vanduren was killed in the International Mine in 1909, his widow filed a compensation claim. The claim contains a letter sent by Davies to the Compensation Board a full nineteen months later:

> Last February I directed the funeral of a man named Felix Vanduren who was killed by a fall of rock in the ICCC [sic] mine. Now I have waited as long or longer as I can afford to wait and I would like to get this matter settled up but I do not wish to press the widow of the deceased as I believe she is rather badly off at present but is willing to pay as soon as she receives her compensation.[117]

Circumstances did not change over the years. The compensation claim files for both Mike Knish (died 1914) and Samuel Edwards (died 1918) both contain unpaid bills from Davies.[118] Davies' successor A.E. Ferguson also had considerable trouble in sometimes having his bills paid. In 1924 Ferguson sent a bill for $238 to the Coleman Local of the United Mine Workers of America, of which they were only willing to pay $120.[119]

Davies remained the undertaker for the Alberta side of the Crowsnest Pass throughout many of the major disasters, including Bellevue in 1910 and Hillcrest in 1914. Eventually, Davies sold his practice to A.E. Ferguson in 1921, a transition that marks a further phase in the professionalization of the undertaking trade. Although Davies was proficient in the art and science of embalming, there is no indication that he had learned it at an accredited school. Instead, Davies' experience probably resulted from an apprenticeship and considerable hands-on practice. Ferguson, however, initially took pains to advertise himself as a "graduate of Worsham College of Anatomy and Embalming, Chicago." Along with these impressive credentials, Ferguson also made the transition from the undertaker-as-businessman to a proprietor of an actual registered business. Ferguson created the first "business" name for undertaking in the Pass when he registered his new enterprise as "Crows Nest Undertaking Co."[120]

Earlier, the undertakers in Fernie had also sought to impress their clients with their qualifications. When William Scott had partnered with Owen Ross in 1905 they advertised in a straightforward manner as "Undertakers and Embalmers." Upon the dissolution of the partnership in 1908, Scott apparently sought to enhance his professionalism and his standing as an undertaker. The ad announcing the establishment of his new business with himself as the sole proprietor is now careful to mention that he has become a "Charter Member of the Manitoba and Alberta Undertakers' & Embalmers' Association."[121]

Throughout his long career as the principal undertaker in the Pass, T.W. Davies also adhered strictly to the professional requirements of his trade. Although it wasn't until the 1920s that undertakers achieved the role of funeral professional,[122] from his very first foray into the business, Davies fostered the theoretical ideal through the

3 Convention of Alberta Undertakers in Edmonton, 1907.
Provincial Archives of Alberta, B7305.

DEATH 63

reality of his practice. He seemed to instinctively understand several of the main requirements of a professional, following in the footsteps of William Scott in Fernie. One of these requirements involved belonging to a professional organization that offered education and training,[123] preferably an association with an acknowledged code of conduct.[124] Davies understood immediately what would later become a basic requirement of his chosen profession: that having a complete knowledge of the industry, and appearing self-confident in that knowledge, would be a desirable characteristic for first attracting and then contracting for service.[125] The funeral services practitioners in Alberta first organized as a professional organization in 1907 under the title of the Alberta Undertakers' and Embalmers' Association.[126] During his first full year as an undertaker, Davies attended the Association's annual convention in 1910 as one of the growing handful of neo-professionals in Alberta[127] (see Photo 3).

Davies also understood the business end of the undertaking trade. While conducting himself with professional decorum, he was not averse to attracting business in other ways. An active man in the community, Davies served as fire chief and as a member of the Board of Trade besides being a successful contractor.[128] His counterpart in Fernie did Davies one better. George "Curly" Thomson had come to Fernie in 1907 and eventually became the sole proprietor of a "first class undertaking establishment." Thomson was also extremely active in civic affairs, taking successful roles as coroner and alderman. In 1918 he crowned his career in public office by serving as mayor of the city, gratis.[129]

While these additional services and offices would have aided both Davies' and Thomson's contacts in the community, Davies also managed to find other methods of raising his profile. One of the more important was the inclusion of his name and services in the funeral notices.[130] The 1908 newspaper accounts of Crowsnest Pass funerals do not mention Davies' predecessor, R. Addison, by name. However, starting in 1909 when Davies took over as undertaker, there was a flurry of statements concerning his involvement with the funerals. At first, the newspapers noted only a short statement of his services: "Undertaker Davies prepared the body for burial." Although these notices of participation remained brief, the scope of Davies'

4   Hearse and Driver (probably Joe Plante, T. W. Davies' assistant), Connor Tiberg's Funeral, c. 1916. Crowsnest Museum, CM-BL-13-12.

involvement in the funeral direction gradually broadened until: "T.W. Davies, funeral director had charge of the funeral" or "T.W. Davies conducted the funeral arrangements."

Davies also took care to continually make mention of any changes or additions to his business. In 1910, the enterprising undertaker added an elaborately decorated hearse to the services he provided. In one of the first ads mentioning the hearse, he very discreetly stated:

> F.W. [sic] Davies has recently procured a hearse and should the residents of Blairmore and Frank have occasion to use the same, Mr. Davies will be pleased to offer himself and hearse at a reasonable figure.[131]

By 1910, Davies had taken direction from William Scott's initial advertisements of two years earlier to use the power of images as a drawing card. Prominently displayed in all his front-page ads was an image of what Davies unabashedly advertised as a "splendid hearse" (see Photo 4).

The undertakers provided a number of services to the residents of the Pass. One of the more important was the ordering and stocking of a full array of caskets according to the needs of their clientele. Davies supplied a full variety of these, ranging from the extremely simple

DEATH   65

5   Women with Casket, c. 1928. Glenbow Archives, NC-54-1902.

DEATH

6 Man in Open Casket, c. 1928. Glenbow Archives, NC-54-836

(Photo 5) that did not even contain handles but had to be carried by ropes slung over the bearer's shoulders, to the ornately carved and decorated (see Photo 6). Undertakers also supplied funeral attendants and other services and goods to the grieving families. When Henry Munkwitz died in Fernie in 1921, the list of services provided by J.P. Lowe the undertaker included gloves, a hearse, three cars, wagon deliveries, opening the grave, telegrams, and phones. That same year when another Fernie miner, William Cole, was buried, the list of services included candelabra, death notices, flowers, a tie, and underwear.[132] When necessary, undertakers arranged for the transportation of the body from the house to the church to the cemetery and conducted funeral services at the graveside. They were also responsible for preserving the body in such a way that, if required, the body could be shipped back home (see Photo 7).

Apart from these administrative services, each of the undertakers also provided comfort and aid to the bereaved. Perhaps the most difficult but also the most important responsibility of an undertaker was to somehow retrieve the dead from the often brutal aspects of their dying and return them to their families in some semblance of their past appearance. In numerous instances, the newspapers related stories of drowning, train accidents, suicides, and the always present mine accidents. In many of these cases, the articles mention that the remains were "taken in charge by Undertaker Davies" or "the body was taken to Coleman by Undertaker Davies and prepared for burial." Though the articles are too discreet to detail the "preparations," Davies was undoubtedly attempting to repair the damage caused by death and to send the body back to the family in a presentable form.

The community could not have needed Davies' professionalism more than when mining disasters struck the Pass. During these times of great stress, Davies displayed the other requirements of a true professional in great quantity: demonstrable skill, integrity of conduct, and provision of a service for the public good.[133] There could be no more horrendous task required of an undertaker than the preparation of scores or even hundreds of bodies and the attempt to repair the worst effects of mine explosions. In these cases, Davies would have recognized that both by his professional and personal conduct, and more importantly by virtue of his specialized role as

7 Funeral gathering at train station, c. 1916. Glenbow Archives, NC-54-2770

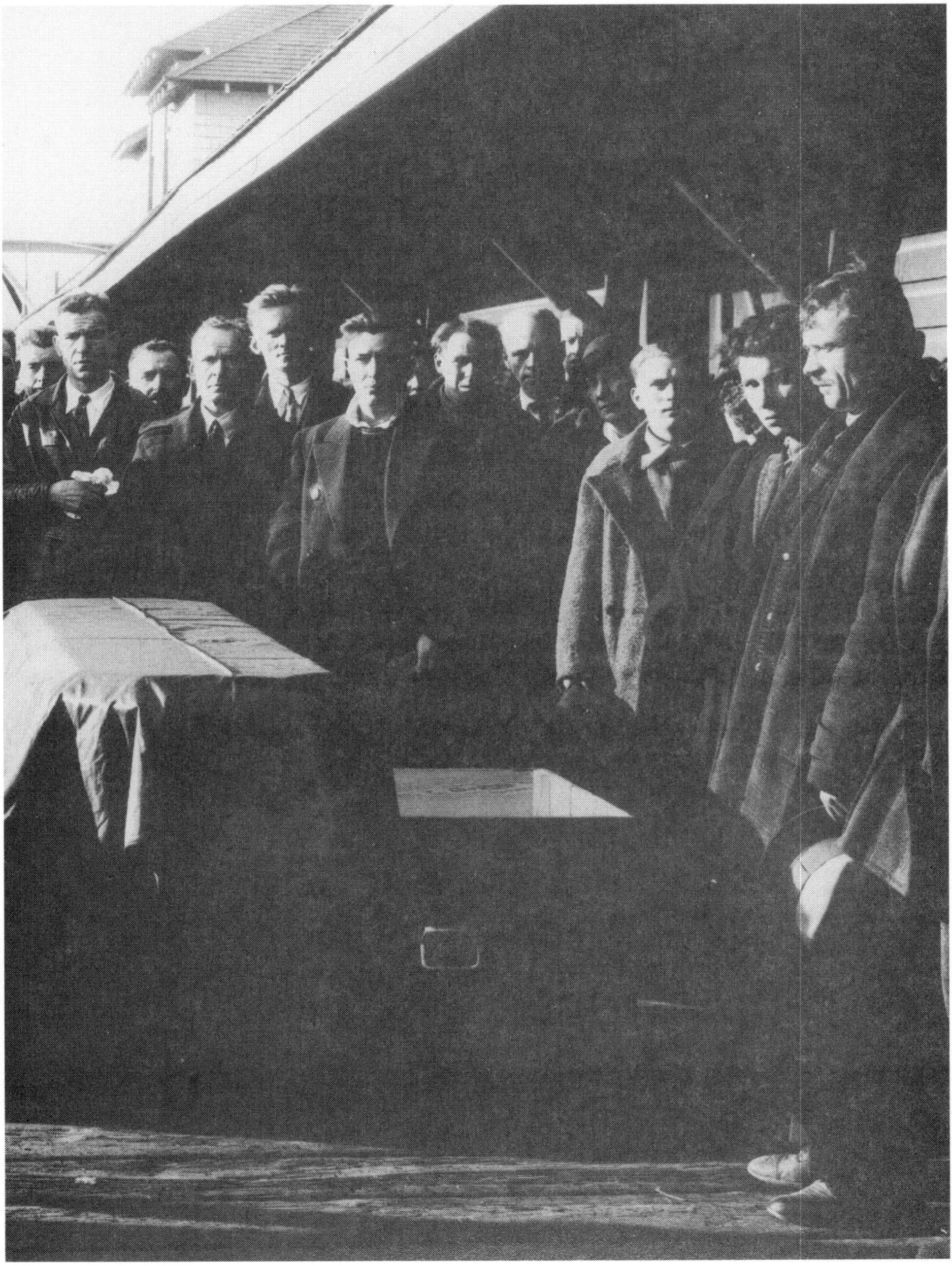

a handler of the dead, his demeanour would go far to help ease the psychological distress of those left behind to mourn.[134]

In the case of major disasters, more than one such professional was often required. At Monongah in West Virginia, six undertakers took charge of the 361 dead in 1907; at Stag Canyon, New Mexico in 1913, nine undertakers were needed to attend the 263 dead. In the Crowsnest Pass, the Hillcrest Mine explosion in 1914 resulted in 189 dead. Three hours after the initial blast, the General Manager John Brown placed a call to Davies requesting the undertaker assume complete control in preparing the dead for burial. Davies immediately placed express orders for "a carload of coffins and caskets from Winnipeg and another from Calgary." These coffins would supplement the stock that the carpenters from the mines could supply.[135] Davies also had to arrange for the transportation of the coffins from the train yard to the improvised morgue in the Miner's Hall. All available wagons, drays, and drivers were pressed into service, with any spare men available to drive them.

With supplies taken care of, Davies then called upon five of his fellow undertakers from Lethbridge, Macleod, Pincher Creek, Hosmer, and Cranbrook. Several of these men also arrived with a number of coffins since the shipments from Calgary and Winnipeg and from the mines would be insufficient to meet the demands. Under Davies' able direction, despite the fact that he was not quite recovered from a bout of pneumonia, these five men worked for six days preparing all the bodies for burial. Some of the bodies were also embalmed. As a present-day undertaker has remarked, the retrieval of the dead from the power that killed them can greatly ease the minds of those left behind.[136]

After such a particularly shocking occurrence, the community placed a vital importance on having the final preparation of the dead for burial accomplished with discretion and professionalism. The quiet efficiency with which this massive task was accomplished, along with the professional demeanour of the men involved, may have relieved in some small way the grief of the community witnessing their work. The fact that these thankless tasks were carried out with complete understanding and respect at Hillcrest is evidenced by the praise lavished on those who worked through those dreadful hours. Two policemen were in charge of the initial washing and identification of the men as they came out of the mine. After the police had accomplished their "arduous as well as grievous task" the bodies were transported to the Miner's Hall that had been set up as a temporary morgue.[137] Here the undertakers fulfilled their role:

> The work of preparing ... bodies for burial was a task that might rightly be described as Herculean, but it was accomplished and every detail carried out without the slightest hitch or bungle.[138]

The writer also evinced considerable faith in the aptitude of Davies when the body count was called into question: "It is hardly possible that the undertaker would make a miscount of the bodies passing through his hands."[139] Davies was in full charge of the burial arrangements, placing each man in a coffin set in a long row according to denomination so that the community services could go forward without any problems.[140]

Davies and his assistant, Joe Plante, never received any remuneration for the work they did on the bodies at Hillcrest.[141] Nor is it likely that Davies would have asked for any. In providing this service, Davies responded with complete professionalism to a desperate need in his community. His response formed an integral part of a much deeper reaction by the rest of the community to a disaster situation.

3 DISASTER

Oh, what a dreadful spectacle for to behold the dead,
When all their lifeless bodies upon the bank was laid. [sic]
To recognize their features, it scarcely could be done,
The uncle from the cousin, or the father from the son.

They were so mutilated, alas! It is well known,
How some of these poor colliers were all to pieces blown,
The head from off the shoulders was severed I declare,
While bodies, legs and arms, they were scattered here and there.[1]

THE EVENT

What is a Disaster?

The debate over what constitutes a "disaster" has been long and varied. Sociologists, psychologists, geologists, anthropologists, and historians have all weighed in with different perspectives. Some of the questions raised include: Does a disaster exist as an event in itself or is it defined by its social consequences? Do disasters cause the effects or are the effects themselves the disaster? Can disasters be defined as events that are "implicit or explicit catalysts for collective action?" Or should the question remain basic and focus on what a disaster itself is, and not on how a society acts and reacts under the conditions of a disaster?[2]

Although the death of a single miner would constitute a disaster for his family, the United States Bureau of Mines has defined mining disasters as accidents or chains of events that result in more than five deaths.[3] Mining disasters were physical agents that created (or presupposed) the need for certain responses on the part of individuals and communities to the aftermath of the events. These events inspired periods of "collective trauma";[4] in communities of single industry like the Crowsnest Pass, everyone in town was either directly or indirectly affected in a significant way (See Table 1 for a list of major disasters in the Crowsnest Pass.)

Mining disasters are somewhat unique when compared to other types of disasters, and as such they require different parameters and caveats. Mining disasters belong to the specific category of technological or industrial disasters to differentiate them from natural disasters (tornadoes, floods, earthquakes, etc.) or human disasters (terrorism and war).[5] Mining disasters place only the miners themselves in direct danger, creating a specific and distinct group of primary victims[6] An explosion in a mine creates a far different effect than a forest fire, earthquake, or flood. These types of natural disasters threaten and can often destroy the entire infrastructure of a community; they tend to "obliterate everything that [gives] people their identity and their lives meaning."[7] When an explosion occurs in a mine, no property damage results other than to the mine itself. Nor does the survivor's physical environment change in any significant way.

There are some problems defining disasters within the parameters of an "event." This definition tends to encourage the idea that the communities experiencing the disaster appear as passive victims.[8] If mining disasters were "events" characterized as "non-routine, destabilizing and the cause of disorder and uncertainty," then this presupposes that the community before the disaster situation existed in complete equilibrium and normalcy.[9] This assumption would be incorrect in the case of the Crowsnest Pass. On some level, every miner and his family were aware of both the dangers present in the mines and the real possibility of death that might result from working there. The residents of the Crowsnest Pass had already responded in some way to the possibility of a major explosion because they lived in a potentially dangerous area. Everyone in the community had individually either experienced first-hand the effects of a mining accident or knew of a neighbour or acquaintance who had lost a loved one. The community infrastructure as a whole had already responded to the threat of disaster given the numerous networks of support that existed. However, while the knowledge of danger and the threat of disaster might have led to a certain expectation of grief on the part of Crowsnest Pass residents, the loss of scores of men would be impossible to anticipate and the effects on the community cannot be minimized.[10]

The Survivors

Mining deaths were sudden. Unlike death caused by old age or long illness, survivors rarely had the opportunity to prepare themselves for the news that a healthy miner in the prime of life had died.[11] And since mining deaths occurred underground, family members seldom witnessed the events that killed their kin. Mining deaths therefore held a certain measure of unreality about them for those left behind. These factors made notification of the living an extremely traumatic experience for all those involved. When accidents happened, the news was usually broken to the family by a clergyman or by friends. Mrs.

Wilson of Pincher Creek had the news of her husband's death in 1908 "gently broken to her by friends."[12] The Reverend Mr. Cook "found a heartbreaking duty to perform as he joined other Pass ministers in visiting and assisting relatives of the 189 men who lost their lives in the [1914] Hillcrest mine explosion."[13]

In more disastrous situations, mining communities had an early warning system when they heard the unexpected blowing of the pit whistle. Julia Makin remembered that although she never heard the actual sound of the Hillcrest explosion in 1914, at the sound of the shrill pit whistle, she and her classmates all knew immediately that something had gone terribly wrong.[14] Short bursts signified the end of a shift, but "the long cry of the whistle ... usually signaled a mine accident."[15] In the case of Coal Creek, where many of the miners lived five kilometres away in Fernie, the first news of a disaster would have to rely on either telegraph signals or the phone line. These worked remarkably well, even in the early years; when the 1902 explosion occurred, a train with three doctors and two coach loads of volunteers only took twelve minutes to reach the scene of the disaster.[16] Only rarely did news of an accident or disaster come as a complete surprise to family members affected. Kay Yates remembered at the time of the 1926 McGillivray mine explosion that she first heard the news when she went out to the store. Only then did she realize that her husband hadn't come home that night.[17]

Many wives commented on their daily exposure to the fear of the potential dangers in the Crowsnest Pass mines. The wife of one McGillivray Creek miner remembered: "It was terrible, they went off in the morning and you never knew if they would come home in the evening."[18] Other miners' wives regularly had nightmares of a hearse showing up at the door.[19] However, a vast gulf exists between *expecting* a loved one's death and *experiencing* that death. To be told that someone you had last seen only a few hours ago and who was very much alive and well was now suddenly very dead creates the necessity for some major adjustments on the part of individuals.[20] As one witness to the 1958 Springhill mine disaster stated: "Telling someone there is no hope is the most terrible message anyone can ever deliver."[21]

Survivor reactions could be expected to range from "disbelief, shock, grief, and dismay to disorganization, hostility, and fear."[22]

Initially, many families dismissed the idea that their husbands, fathers, sons, and brothers were gone. Shortly after the 1902 Coal Creek explosion, "there was at first incredulity as to the truth of it, which was then quickly succeeded by wild excitement from the relatives of those who were supposed to be working in the mines at the time." Most of the wives reacted predictably as well in that they would not believe the horrendous news until "undeniable proof in the shape of recovered bodies had arrived."[23]

A natural response of survivors to sudden death manifested itself in the families of the dead converging on the scene of the accident. They gathered together to gain courage from one another as they waited for news and for the incontrovertible evidence that their loved ones were indeed gone. The most precious commodity for those at a disaster scene is accurate and timely information.[24] Mining disasters tended to compound this need for information, since the incidents happened underground where no one could be immediately sure of what had actually occurred or the extent of the damage. Mine officials themselves seldom had accurate information, so rumours could run rampant. The newspapers initially reported at least 150 dead at Coal Creek in 1902, although the eventual total of 128 was dreadful enough. The *Fernie Free Press* also found that the Company clerk would not give out any information after the 1917 explosion that killed thirty-four men, so the reporters questioned the crowd for information.[25] The not-knowing was terrifying and disconcerting for those waiting at the mine mouth. As one modern undertaker has stated: "It is why MIA is more painful than DOA ... knowing is better than not knowing."[26]

Identifying the Dead

Evidence or confirmation of actual death could be a long time coming given the problems with identifying the dead after a mine explosion. Every miner dressed in more or less the same manner and carried identical equipment (see Photo 8). However, each miner had two claim checks, or brass disks, issued to him with an individual number stamped on them for the purposes of tracking and identifying the miners in the event of an accident. The men would hand in one of

8   Miners posing with Davey Safety Lamps, 1910.
    Provincial Archives of Alberta, A2038.

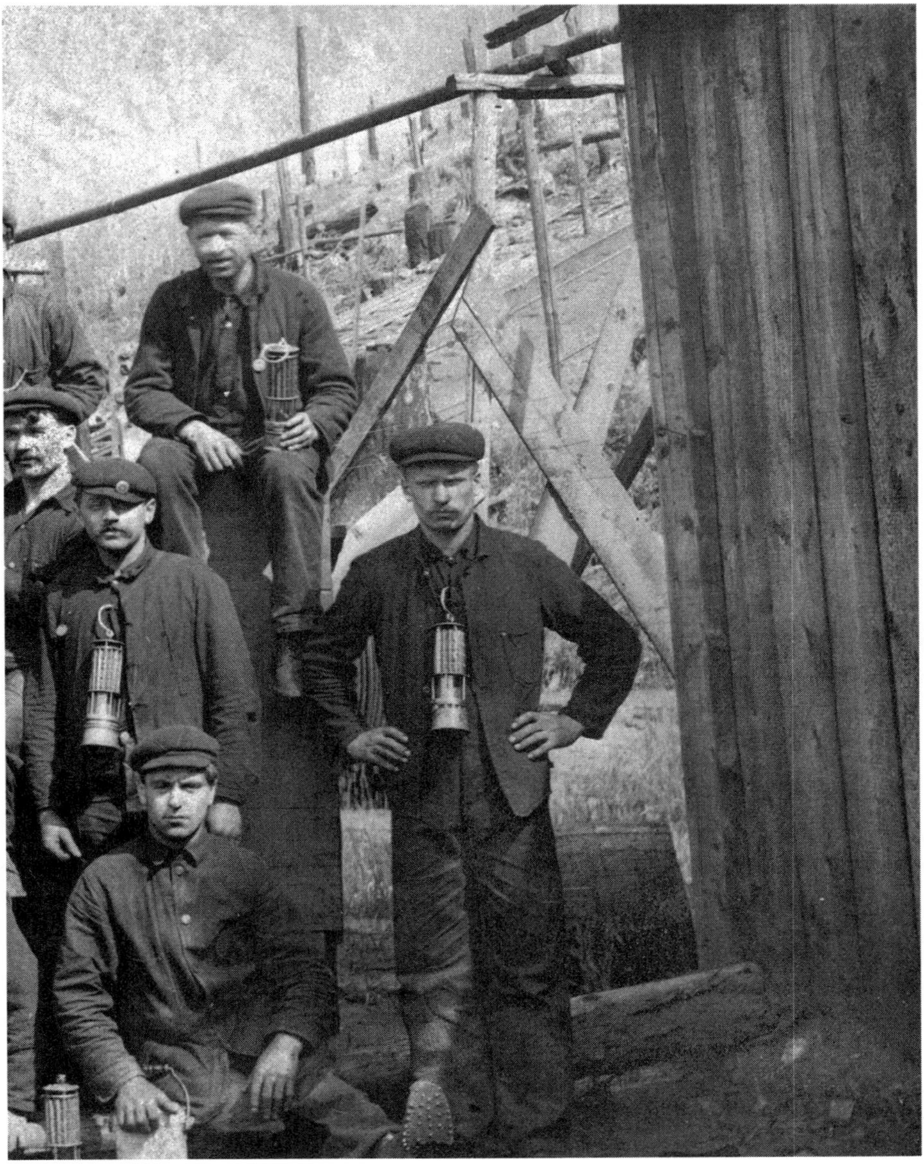

the disks to the timekeeper, who would hang it on a board to indicate that the miner had actually gone into the mine. The miners kept the second claim check on their person, either in their pockets or hung around their necks. The number on the claim check would serve to identify them in the case of a disfiguring accident.[27] A year after the 1902 Coal Creek explosion, the bodies of two of the men were finally discovered inside the mine: "There was no trouble in identifying the men by the numbers of their car checks."[28]

However, any number of problems could arise because of the shortcomings of this identification system. The timekeeper died in the 1902 Coal Creek explosion, making it impossible to obtain more than a partial list of those who were still down in the mine.[29] At the Hillcrest disaster in 1914, the timekeeper turned away two miners who showed up drunk, but then absent-mindedly hung their claim checks on the board anyway.[30] Their families at first naturally assumed the men were in the mine with the rest of those presumed dead. Another family experienced similar anguish when a later explosion killed William Powell in the Bellevue mine. Powell had for some unknown reason been carrying the claim check of his stepfather William Hadwell. The mine manager at Bellevue later reported: "Please note that my report ... affecting William Hadwell who was working under check 339 ... at the inquest it was brought out the correct name of this worker was William James Powell."[31] No explanation for the mix-up in claim checks was ever given.

Aside from confusion caused by administrative problems, the manner in which men died often made identification of their remains extremely difficult. Explosions could char the men beyond recognition, tear off limbs, crush identifiable features, or result in decapitation. Clothes could be burnt or torn off by the blast and any claim check subsequently lost in the debris. A description of the bodies at the Hillcrest mine explosion illustrated both the monumental task and the painstaking detail necessary in identifying the remains. The description of one body read: "Unknown—gold second left upper incisor, fractured skull, arm and leg, burns on head and body." Another body was portrayed more simply: "Unknown—head off, burns."[32] If the recovery efforts took longer than a few days, often the bodies deteriorated to such an extent that identification was impossible. Following the Michel explosion in 1916 that killed eleven men, one body was so badly putrefied the doctor could not make a positive identification.[33]

More sophisticated identification procedures were actually known during this period. Anthropologist Alphonse Bertillon devised the "Bertillon System" of identifying individuals in 1882 by taking detailed measurements of the body. In the 1890's Sir Francis Galton had developed the fingerprinting method. However, although these techniques were known, miners in the Crowsnest Pass were neither measured nor fingerprinted as part of their employee record. As a result, identification devolved to one of two far more prosaic methods: identification of personal property or body parts.[34]

In the case of personal property, sometimes clues to the identity of the dead could be found on their clothing:

> But when the corpse is viewed, there's none can claim,
> Nor yet the mangled body name.
> Some buttons they take from the clothes,
> And ask if anyone these buttons knows.
>
> The buttons they are passed around,
> And soon an owner there is found.
> A mother cries: "Those are my son's!"
> To get him in her arms she runs.[35]

Similarly, after the 1891 Springhill Nova Scotia explosion, a mother identified her son by recognizing her own stitching on his clothing.[36] At other times, the inventory of personal effects, such as jewellery, rings, pins or identifiable embroidery on handkerchiefs could lead to a correct identification.[37] Some miners took special care to wear jewellery that could identify them in the case of a disfiguring accident. Sydney Hutcheson witnessed the charred remains of men caught in a 1917 Coal Creek mine accident. When only those men who wore rings could be accurately identified, Hutcheson resolved to purchase and continually wear a ring himself.[38]

When personal property failed to ensure an accurate identification, distinguishing bodily characteristics could provide some clues. The sister of one miner killed in the 1902 Coal Creek disaster finally identified her brother after mentioning that he had two missing fingers. However, the recovery teams still had to find the arm with the missing fingers before identification could be completed.[39] The recovery team declared William Machin killed in a powerhouse explosion at the International Coal and Coke Company in December 1922. Witnesses saw him enter the powerhouse just before the explosion, but all that could be found in the debris afterward were charred bones. There was considerable surprise three months later when workmen found Machin's body floating in the Oldman River. His father identified the body correctly due to a childhood scar on Machin's knee. No explanation was ever put forward for the wrongful identification,[40] nor was the owner of the bones found in the powerhouse ever identified.

Even before identification could be made, considerable detective work was involved in finding the bodies of the missing men. Mine rescuers and recovery teams were in desperate need of blueprints of the mine or for mining engineers' schematics. Rescuers needed these documents not just to try to find their own way into the mine, but to understand where the miners were working at the time of the explosion.[41] These documents were often misleading, however, for a number of reasons. In the case of the miners found one year after the 1902 Coal Creek disaster, these men could not initially be found because their supervisor had changed their working place on the day of the explosion. The supervisor had also been killed in the explosion.[42] Also, when miners tried to escape from an explosion, they would naturally move away from their working places in an effort to reach the surface. Their flight would make it more difficult to predict who was where, or whose body parts were found at different places in the mine. During the 1914 Hillcrest mine explosion, "only the men killed instantly were at their stations. The others had swarmed through the passages, vainly trying to find a way out of the deadly labyrinth."[43]

A further complication could arise when men not scheduled to be in the mine were present at the time of the explosion. In the Monongah, West Virginia explosion of 1907, a salesman died with the miners. He had been going from room to room selling life insurance.[44] Repairmen who made unscheduled visits to the mine were also sometimes trapped or killed by an explosion. Rescuers took some considerable time after the 1926 McGillivray mine explosion trying to find a repairman named Johnson "whose likely whereabouts could not be told."[45] They did not discover Johnson's body until some days later.

The intimacy involved in handling and identifying the dead caused considerable anguish to the mine recovery teams and to those attempting to reassemble or clean the bodies. During the Hillcrest mine disaster, two policemen were assigned the unenviable task of finding the claim checks on the bodies and taking charge of any personal property that might lead to identification. The police corporals later cleaned and gathered together various body parts in the washhouse. Their supervising officer did what he could to help them endure their difficult work: "The whisky bottle on the shelf in the washhouse ... continued to reproduce discreetly and promptly."[46] Their supervisor later commended the two corporals for their work and officially recognized them for their unwavering service which entailed "the work of handling the mutilated and battered bodies ... [it] was a gruesome job and I believe the men who did this work deserve recognition for their services."[47]

If neither the rescuers nor the recovery teams could make an immediate identification of the dead, then the process devolved to the already traumatized relatives. The wives and families of those killed in the 1910 Bellevue mine explosion waited patiently by the hundreds outside the washhouse before being allowed inside to view the remains.[48] The families at the 1914 Hillcrest mine were not initially as stoic: "Every time the coal cars came up from the mine with the bodies wrapped up in blankets, the women would run to the cars and the Mounted Police were there in their red uniforms, keeping them back." Eventually, however, the families were allowed to claim their dead. Even children were not exempt from the need to see the bodies of their fathers, uncles, or brothers. Although only eight years old at the time of 1914 Hillcrest mine disaster, Julia Makin remembered the caskets lined up for the funerals: "Every time someone opened one [casket] up, I kept running up to see my dad, but I never did."[49] Julia was not alone in her need to see the dead one final time. A photograph of the Hillcrest mine disaster shows the coffins set out in a row and family members lifting the lids (see Photo 9).

During the first part of the twentieth century, the bereaved followed the common practice of viewing the deceased.[50] Sydney Hutcheson remembered that in the aftermath of the 1917 Coal Creek explosion, "the bodies were badly burned and anyone could go into the [undertaker's] parlor to look. Every day a group of us checked after school."[51] This could be construed as one part morbid curiosity, but the practice also reinforced the certainty of death when a person could actually witness and confirm the results. There were some instances in the Crowsnest Pass when family members were denied a last look at the mangled remains of their loved ones. Many of the more broken bodies were kept inside the Hillcrest mine until after dark to prevent the families from seeing them.[52]

However, despite this apparent sensitivity, many family members still needed to take one last look at their loved ones before burying them. When a bereaved mother walked thirty-five miles to Fernie in 1902, "her sole remaining hope [was] to dress the shattered body of her remaining boy."[53] She believed that her parental duty would not be complete until she had buried her child.[54] The effects of not being able to see a loved one for the last time are obvious from the account of one widow at the Springhill mine disaster of 1958. She remained convinced for the rest of her life that the coffin held only rocks and stones and not the remains of her husband.[55] In such cases, final contact with the dead was not solely for purposes of identification. A final view of the deceased also performed the vital function and "long standing right of the living to declare the dead, dead."[56]

THE RESPONSE

Grief and Shock

The sudden loss of a loved one, or in the case of mining disasters, the death of many members of a community, produced understandable grief reactions in individuals. Symptoms of acute grief could comprise: somatic distress including exhaustion, shortness of breath, and the need for continual sighing, preoccupation with the image of the deceased, guilt, hostile reactions, and loss of patterns of conduct. In order to work through these symptoms, those left behind could ease their trauma through "grief work," or "emancipation from bondage to the deceased, readjustment to the environment in which the deceased is missing, and the formation of new relationships."[57] A report concerning Isabella Petrie, who lost three sons in the 1914 Hillcrest mine disaster, described her as "lying in the hospital [with] doctors doubtful of her recovery."[58] Although she did recover enough to live another eleven years, Mrs. Petrie typifies a person who could not follow through with the necessary "grief work" to move forward past the tragedy:

> The wound would never heal. Mrs Petrie recalled an experience of 1906. She told of a terrific storm in the Atlantic as they were crossing to Canada. The boat had been caught in a whirlpool and spun around and around and around. June 19, 1914 [date of the Hillcrest mine disaster] Mrs. Petrie wished she and her eight children had been sucked into the whirlpool together.[59]

There is also another brief mention in the report of an unidentified woman who "went insane leaving her three children to the mercy of the world."[60]

Unfortunately, these accounts are the only two available that describe in any detail individual grief reactions. Determining a trend or pattern in the Pass is therefore difficult given the dearth of recorded individual reactions. However, the initial grief and the resulting actions of individuals were well documented by the press. The newspapers recorded the first rush toward the mine entrance and

9   Mourners with coffins after mine disaster, 1914.
    Glenbow Archives, NA-3965-69

88   DANGER, DEATH AND DISASTER

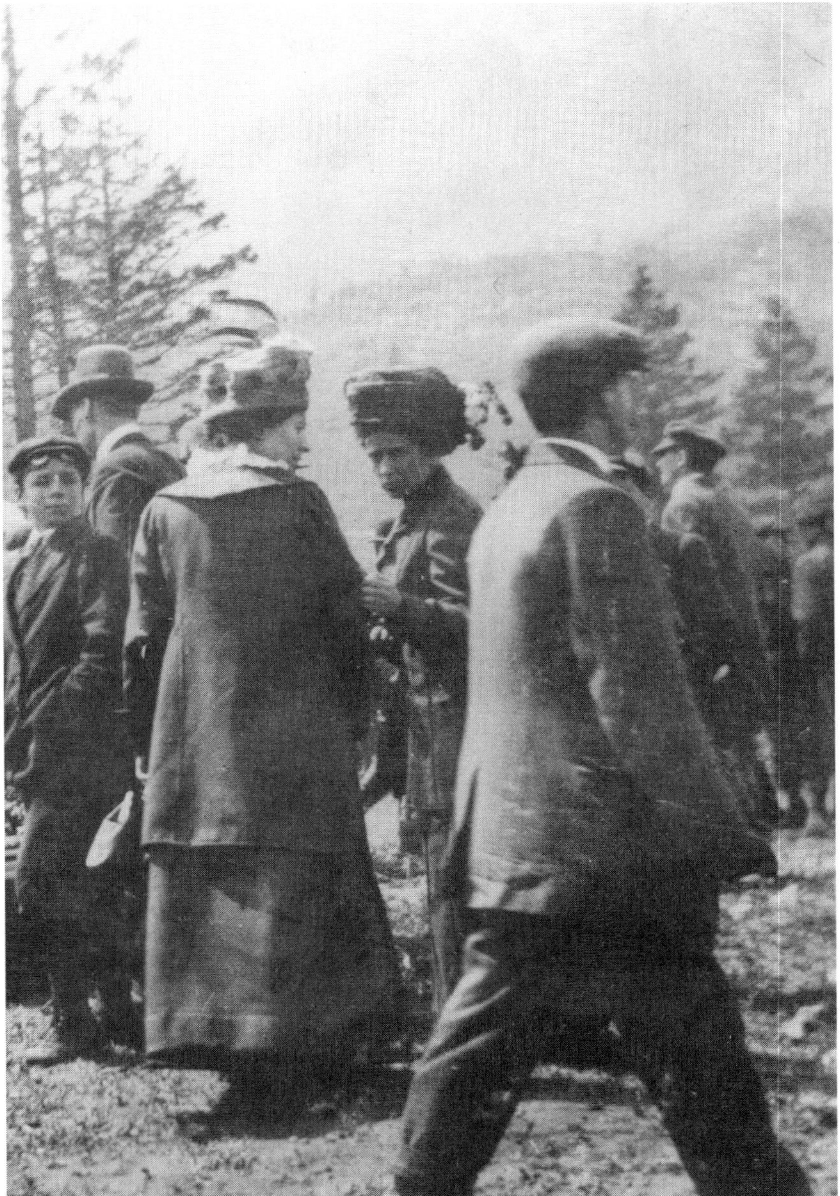

the sometimes excessive zeal of the family members trying to reach their loved ones, or to gain any scrap of news. Also documented were scenes of women later standing quietly together "eyes red with weeping, holding close to each other or vainly trying to comfort children that clung to their aprons."[61] An editorial following the 1926 McGillivray mine explosion attempted to put the initial grief reaction into words following the death of ten men:

> The shock sustained by the community, was indeed, a very severe one, but how infinitesimal compared to the blow suffered by the stricken wife, the bereaved mother and father, brothers, sisters, and fatherless children? ... A puff of smoke, and in a few fleeting seconds ten homes are made desolate.[62]

A parallel could also be drawn between the thoughts of Crowsnest Pass families to those voiced by more recent victims. People who survived the flood caused by a breach in the mine till at Buffalo Creek in 1972 remembered: "You can't think straight ... I don't know what to do ... nothing will ever be the same again."[63] Similarly, family members at Coal Creek in 1902, waiting for the updated lists of miners brought from the mine, found themselves "consign[ed] to the depths of despair."[64] Women waiting for word during the Hillcrest mine disaster were described as "terror-stricken and heart-broken."[65] Margaret Dunlop's diary entry following the 1926 McGillivray mine disaster notes: "What a day of panic and horror! My thoughts are in a jumble."[66] The men who had died represented key persons in the community of the Crowsnest Pass, and as a result the family members were experiencing a "disintegration of their social system."[67]

The chance always existed that some men would remain alive but trapped following a major explosion. They might find themselves cut off from escape by a coalfall, or their passage out closed to them due to fire, debris blockage, or an impenetrable wall of blackdamp. In the latter situation, they would probably retreat further into the mine, barricade themselves in a room and wait for help. Miners all knew that if they were denied the mercy of instant death, and they somehow managed to evade the blackdamp, that their time would be spent in the "suspense and the stillness and the waiting, waiting, waiting."[68]

Most miners had an understandable horror of being left to die slowly, trapped in a mine. They knew entirely too well their inevitable journey to the end, either from personal experience of being trapped before, accounts from their friends, stories from the newspapers, or from popular culture. In 1869, the Avondale mine disaster had inspired a ballad that eventually became one of the most widely known and sung throughout the mining world. This was in part because its author "made such a vivid word-picture of the horror of one hundred and eight men and boys trapped by fire deep in the earth dying from slow suffocation."[69] Trapped miners were only too aware they could also die excruciatingly slowly from thirst and starvation. Miners cut off from escape sometimes ate the bark off the timber props, sucked on coal to get their saliva flowing, collected seepage from the gutters or the walls, or drank their urine.[70]

Mostly, they sat and waited. In the Cherry, Illinois mine explosion of 1909, men passed the time by writing down their thoughts. Joe Pigati, who made it out alive after seven days underground, shared the thoughts he had written to his wife:

> I am writing in the dark because we have been eating the wax from our safety lamps, I also have eaten a plug of tobacco, some bark, and some of my shoe. I could only chew it. I'm not afraid to die. O Holy Virgin have mercy on me ... It has been very quiet down here. Goodbye until the heavens shall bring us together again.[71]

Other diaries were found written by men not as fortunate as Pigati. Samuel D. Howard left a short commentary of his thoughts written over the long hours that he struggled to remain alive:

> 4 p.m. following day – dying for want of air.
> 2 a.m. – still alive. We are cold, hungry, weak, sick and everything else.
> 9:25 a.m. – still breathing. Something had better turn up or we will soon be gone.
> 16 to 1 p.m. – the lives are going out. I think it is our last. We are getting weak.[72]

Howard died of blackdamp after waiting two days for rescue.

For those who made it out of the mine alive, the experience of being trapped in the dark seldom left them. In 1963, eleven miners were rescued from a German mine after fourteen days underground, during which time ten of their colleagues died beside them.[73] In 1958, in Springhill, Nova Scotia, nineteen miners survived, twelve after six days underground, and the remaining seven were found alive after eight and half days.[74] The symptoms exhibited later by the survivors included one already identified as an acute response to grief: "intense preoccupation with the image of the deceased."[75] After ten years, one of the German miners admitted to having hallucinations of his dead colleagues peering around corners at him in the mine. Several of the German miners were also still experiencing "somatic distress" in the forms of nightmares combined with feelings of guilt. One miner, whose friend drowned when they could not maintain contact with each other, continually had an image of rushing water and heard his friend asking, "Why didn't you hold me tight?" Gorley Kempt, one of the Springhill miners, experienced periods of light phobia after the 1958 explosion. His wife described Kempt's experience:

> When he came out ... it was not unusual for me to come home and the house would be in darkness and him sitting in a corner. Yes his nerves. Maybe he kept it all bottled up and after he was out his nerves went then."[76]

Joe McDonald, who was trapped in both the 1956 and the 1958 Springhill explosions, could not bear to sleep without a night light.[77]

There are virtually no records of the reactions of men who survived explosions in the Crowsnest Pass. Only the words of Frank Bombadier, who was trapped under a fall of coal for twenty-five hours in 1930, survive to give some indication of what he went through. Although eventually rescued, Bombadier found that his partner Frank Cizek had died during the experience. This realization led to some feelings of guilt: "As he lies in the hospital now Bombadier cannot understand how he escaped and poor Cizek was killed."[78] Perhaps Charles Elick had similar nightmares and experiences. Along with seventeen other miners, Elick dug himself out of the Frank mine after being trapped for thirteen hours following the 1903 Turtle Mountain slide.[79] If he did have nightmares in later years, it did not affect Elick's choice of occupation; he would later die in the 1914 Hillcrest mine explosion.[80] Frank Haccart had survived being buried for twenty days after the Courrieres explosion in France that killed 1,099 men. He immigrated later to Canada, where he chose to continue mining at Coal Creek.[81]

Mine disasters touched off grief responses not only in individuals but the entire community. In the case of sudden, mass death, disaster researchers have found that:

> The community's response [to disasters] involve[s] a large segment of the ... population [and that] the organized effort is to some extent a distillation and partial reflection of the community's relationship to the dead.[82]

Given the tight-knit nature of the mining communities in the Crowsnest Pass, community reaction to tragedy in the mines involved an instant rallying around those in need. In general, the Crowsnest Pass communities subscribed to the idea that "we borrowed our strength from one another and repaid it in kind, not knowing if we had any left to share. Regardless of nationality or religion, we were united in hope, fear and prayer."[83]

One of the first responses of a community involved in a disaster situation produced a "convergence" of various groups on the scene of the tragedy. Since mining communities are such that almost every home is linked in some way to the mine, this comes as no surprise. "Convergence behavior" is defined as that "striking aspect of human behavior in disaster [of] the informal, spontaneous movement of people, messages and supplies toward the disaster area." Five general types of groups tend to accumulate at disaster scenes: returnees, the anxious, helpers, the curious, and exploiters.[84] Fortunately, exploiters in the sense of looters were never mentioned in any of the Crowsnest Pass disaster narratives.

When Turtle Mountain slid into the valley at Frank in 1903, burying the entrance to the mine mouth, "returnees" were evident in the case of families and individuals who had by chance spent the night elsewhere and then returned home. A work crew scheduled to join a camp located in the path of the slide ended up delayed that night at a siding at Morrissey; Lillian Clark stayed overnight at the boarding house where she worked and avoided the slide that killed the rest of her family.[85] These individuals, and many others who had property or family in the path of the slide, would have rushed to the area soon after the tragedy.

By far the largest group exhibiting convergence behaviour were the "anxious," the family members of miners trapped in the mines. Wives, sisters, and daughters would have experienced an overwhelming need to reunite with their families. Search behaviour, or family members seeking to find each other, "is usually prominent during impact and the immediate aftermath of the disaster."[86] The newspapers all relate variations of the "women waiting at the gates" behaviour. The women at first exhibited great distress following the initial rush to the mine mouth and later stood waiting quietly for word of the miners' fate.[87] The "curious" were also strongly in evidence during disasters in the Pass. Immediately following the 1903 Frank Slide, his superiors summoned William Pearce, District Mine Inspector, to the area by train. Pearce reported that "a lot of persons attached themselves to the train at Macleod."[88] Following the 1910 Bellevue mine explosion, "hundreds came to view the dead."[89]

An important community response to disasters is evident in the number of "helpers," the fourth group of people converging on the scene. Some of these helpers were initially drawn to the scene solely by curiosity. The rescue of many men and women pulled alive from the Frank Slide can be credited to impromptu rescue parties formed from those first hurrying to the scene out of interest to see what had happened. George Bond stayed in Frank at the hotel the night before the 1903 slide. Immediately following the slide, Bond set off with a group of other men to take a look at the destruction. He and his party ended up rescuing members of the Leitch family and others from their shattered homes.[90] Tom Roach, a trapper and miner who was thrown into the drunk tank on the night of the slide, also found himself joining a rescue party with his fellow detainees. Roach never went back into the mines after the slide.[91]

Other helpers divined what was needed and went out of their way to supply it. William Hutchinson, the engineer at the 1914 Hillcrest mine disaster, remembered: "I practically lived at the mine for days afterwards, sleeping on the floor of the power house and living principally on coffee and sandwiches brought up from the town."[92] The women of the Pass were responsible for providing most of the food to the rescuers. Ethel Price recalled "the women making piles of sandwiches and gallons of tea and coffee, and carrying it up to the mine hill."[93] Julia Makin remembered helping the women carry the food even though she was only eight years old.[94] George Cruickshank tried to alleviate needs when he opened up his general store after the 1914 Hillcrest explosion and "supplied many items required for the emergency situation, including bolts of material for wrapping the bodies."[95] Harry Gate and many others hurried to assist with transportation of the coffins.[96]

Some odd notes of discord should be mentioned. During the 1902 Coal Creek disaster especially, the town split along ethnic lines.[97] Reports in the newspapers commented bitterly on the lack of help from Slavic community: "... they have been conspicuous by their absence from the work of rescue."[98] Whereas apparently the English speaking miners were unstinting in their efforts to assist in the rescue, "the track from the mines to Fernie was lined with foreigners making their way down."[99] In searching for possible explanations for this behaviour, it may be speculated that these men were hurrying to be with their families and to report to their friends and neighbours the news of what had happened. Fortunately, this situation turned out to be an isolated incident in the history of Crowsnest Pass disasters. Human nature may seem to indicate that conditions of severe stress would either create conflict or increase the likelihood of disagreements and quarrels in a community. However, this is not the norm, nor was this behaviour mirrored in the majority of disaster circumstances in the Crowsnest Pass.[100]

There were also isolated incidents of mine rescue workers not being given all possible aid at the time of disasters. During the 1910 Bellevue mine explosion, some of the rescue parties were refused service at the hotel: "... we went up to the hotel and were refused any grub, after we had been working all night. They said that it was out of the regular meal hour, and it made us pretty sore."[101] An editorial following the 1914 Hillcrest explosion also makes mention of the lack of help from certain quarters: "It is regrettable to know that right here in our own community are people who are entirely void of the spirit of hospitality."[102] The writer goes on to outline all the numerous instances of help that were offered, without going into any more detail on what was withheld.

Since these were such isolated instances, the editorial writer in 1914 might simply have taken the opportunity to express a personal grievance, just as the instance of the hotel refusing service in 1910 could have easily been the management's attempt to maintain some sort of order in the face of overwhelming disaster. In fact, the provincial government picked up the tab for the meals supplied during the Hillcrest recovery efforts.[103] In the vast majority of cases, the communities of the Pass responded in all possible ways to assist those in distress in the aftermath of disasters.

Community Response

Mutual Aid Societies and Fraternal Organizations

Some of the most important assistance came from the numerous ethnic organizations in the Pass. Mutual aid societies performed several functions for ethnic groups: they provided some insurance in the case of sickness or death for their members; they gave an outlet and a haven to immigrants who may have been otherwise overwhelmed by the unfamiliarity of their new home; and they fostered a sense of community among those who were displaced from their old surroundings. The three dominant ethnic groups in the Crowsnest Pass were the Italians, the Polish, and the Slavic communities (see Charts 1–3). These communities tended to first locate themselves geographically close to each other, producing the Dagotowns and Slavtowns that were common throughout the Pass. The establishment of ethnic mutual aid societies almost immediately followed. An Italian Mutual Benefit Association formed in Lille and Coleman in 1906 which eventually joined up with its Fernie counterpart that had been established following an increase in the Italian population there. The earliest Slovak societies formed at Fernie in 1900 and Natal in 1903, while the very active Coleman Polish Brotherly Aid Society established itself in 1916. The Finns had also formed their own society by 1910.[104]

In addition, there were many fraternal organizations in the Crowsnest Pass.[105] In Fernie, there was a rush to establish fraternities at the turn of the century: the Ancient Order of Foresters inducted thirty members into their new lodge in 1900; the Masons organized with forty members in 1901, followed the same year by the Woodmen of the World; sixty names were put forward for the Knights of Pythias in 1903; and the Fraternal Order of Eagles had organized by 1904. On the Alberta side of the border, lodges were not far behind, with the Masonic Lodge established in 1906 and the Oddfellows Bellevue Lodge in 1914. These were later followed by various Lodges of the Eagles and the Moose. Nor were women's organizations absent. Women met frequently under such banners as the Women of the Mooseheart Legion, the Pythian Sisters and the Rebekah Lodges. Women also banded together as Ladies Aid societies in both Fernie and Coleman. As one member stated: "Though we were not trained social help, we all knew where help was needed and we were all willing to step in with what resources we had."[106]

Both ethnic mutual aid societies and fraternal organizations responded to two separate needs in the community during disaster situations. First, they fulfilled the pragmatic role of providing material goods, money, and organized relief to the bereaved families of each member. The Oddfellows bylaws provided for a "Widows, Orphans and Education Fund" and also gave provision for a "Nurse and Funeral Tax" should any of their members ever have need for these funds.[107] When F.B. Jackson took ill and died at Frank, the Oddfellows stepped in: "Before his death, the Odd Fellows [sic] visited him and rendered ... aid. After [Jackson] died, the I.O.O.F. took care of the body, saw that it was properly prepared for burial and tenderly placed it on the train en route to Manitoba."[108] The Masons collected charitable funds through annual dues and used the money to help their members in times of sickness and death. During the 1914 Hillcrest mine disaster, the Mason's Sentinel Lodge lost eleven members and contributed several thousand dollars to the relief fund.[109]

These organizations and societies regularly contributed to the funeral expenses of their members. The Italian Mutual Benefit Association's charter declared that in the case of death, $75 would be provided for a funeral, plus a contribution of $1 from each of its members.[110] The Coleman Brotherly Aid Society pledged $100 to a member's widow,[111] while the Oddfellows also pledged at least $30 toward all funeral expenses.[112] The Masons contributed toward their members' funerals as well, sometimes bearing the full cost of the interment and burial.[113] The average cost of an adult funeral toward the end of this period totalled $199.43, although the actual price in the Crowsnest Pass might have been somewhat less.[114]

More importantly, these organizations provided internal support mechanisms to those members and their families experiencing grief, bereavement, or just the effects of dislocation. Walter Chuchla arrived in the Crowsnest Pass in 1926 shortly after the McGillivray mine explosion.[115] Eight Polish miners had been killed, and Chuchla attended the funerals to show his support for the larger community. In return, Chuchla found the funerals a useful network of communication. He soon learned of the Polish Society of Fraternal Aid in Coleman that pledged: "The Society is like a family in which you will find kindliness, friendship, brotherliness, and love."[116] Although the Masons contributed large amounts of money after the 1914 Hillcrest mine disaster, the bereaved community would have appreciated far more the expenditure of the Mason's "personal energy [used] to soften the grief of the families left behind."[117]

Mining Unions

The mining unions of the Crowsnest Pass also assisted the communities in dealing with disasters both large and small. The first mine union in the Pass was the Gladstone Union of Fernie, established in 1899. The Western Federation of Miners took hold in the region at the turn of the century but was soon ousted in 1903 with the establishment of District 18 of the United Mine Workers of America. The miners later flirted briefly with the One Big Union in 1919–20 but then re-formed under the United Mine Workers again in 1921.[118] The constitution of the United Mine Workers established the principles under which they operated:

> ... believing that those whose lot it is to toil within the earth's recesses surrounded by peculiar dangers and deprived of sunlight and pure air producing the commodity that makes possible the world's progress are entitled to protection and the full social value of their product we have formed District 18 United Mine Workers....[119]

While the unions did endeavour to protect the miners in many ways, the presence and assistance of these organizations could be a mixed blessing.

The United Mine Workers fought for the safety of the miners, for fair pay, and for better working conditions. The union investigated when the miners complained about problems in their working areas and when they were dismissed from their jobs. The majority of strike action initiated by the union was for better pay.[120] This may be equated not simply as a struggle for extra money, but as compensation for the dangerous occupation of mining.[121] Miners did appear to make a decent wage, and were often the highest paid industrial force in Western Canada.[122] Miners at Coal Creek in 1903 were paid anywhere from $3–$6 as a daily wage, depending on the amount of coal mined.[123] Assuming that the miners could work relatively steadily for perhaps fifteen to twenty days a month, this could amount to anywhere from $45 to $120 per month. In contrast, a CPR section hand was paid $39 per month, common labourers might make $50 per month, and some clergy were lucky to be paid $29 per month.[124] The police constables at Fernie and Michel made $2.50 per day in 1902.[125]

While the miners appeared to make a good wage, appearances could be deceiving. Although in 1910 a bushel of potatoes was only one dollar and three pounds of beans would set a household back by twenty-five cents, a pair of work boots at full price was a costly $3.50.[126] The mine management and the union had a war of words over the rates of pay during a strike at Coal Creek in 1903. The managers stated that "we do not think that any fair minded coal miner could decently expect a larger average than $4.50 for 8 hours of work" after all deductions had been made. The union countered that the full price of deductions was deceptive, giving as one example the instance of John Spok, who worked nine days at Coal Creek, was paid $20.40 and then had $22.50 deducted from his pay for backhand labour.[127]

The union completed a similar survey of actual wages during the strike of 1911. After working steady for a two-week period, a miner received $13.50 for thirty-six cars of coal. However, after deductions for board, powder, squibs, union dues, library dues, monies paid to the check weigh man, the doctor, the pick sharpener, and the hospital, the miner was in deficit by $3.10.[128] Mining was also never a steady five-day-a-week occupation. Between strike actions, slowdowns in production, and disputes between the managers and the railway when coal cars were unavailable, work in the mines could be sporadic at best. The cost of living was also a constant factor in the fight for extra wages. The average weekly cost for a family budget of food, fuel, lighting, and rent rose from $9.37 in 1900 to $21.56 in 1928.[129]

Strike action was always a double-edged sword. Strike pay could be either inadequate or non-existent. In 1924 the miners received no assistance from the United Mine Workers due to the demands on the international treasury, a situation which led to the formation of the Mine Workers Union of Canada in defiance of the United Mine Workers. Oftentimes the miners returned to work without achieving their goal, and sometimes returned to less pay than when they had started. Strikes could also cause divisions within the union membership, harming their role as a network of support in the community. In several of the strike reports, the men are described as being "not very enthusiastic."[130] Strike action could also result in destruction to the mine through roof falls and damaged roadways that the miners would first have to repair before getting back to the work. When on strike, some miners were desperate enough to find work elsewhere, sometimes with fatal results. Arthur Cartlidge sought work in Calgary during the 1912 strike. He was caught in a cave-in while digging a sewer line and died of his injuries.[131] Peter Blane Anderson chose to work in Mountain Park, a Coal Branch mining town, following the exhaustive strikes of 1922 that shut down production for more than 120 days in certain areas.[132] Anderson died in the mine on 13 November 1922.

The unions did play a significant role in bringing the various disparate groups in the Crowsnest Pass under one common banner. John Mitchell, the president of the International United Mine Workers of America, recognized the ethnic nature of his constituents and used it to gain supporters. He declared in 1899: "The coal you dig isn't Slavish or Polish or Irish. It's just coal."[133] The leaders of District 18 followed Mitchell's lead and went to great lengths to include all the ethnic groups of the Crowsnest Pass within their fold. The

District Ledger, the official mouthpiece of District 18 based in Fernie, published a page of news in different languages for the Italian, Polish, Slavonian, and French brothers.[134] The union also translated their bylaws and literature into Italian and Slavonic, engaged interpreters at meetings for the benefit of the ethnic groups, and chose Italians and Slavs from among their membership to stand in pay lineups and seek out the men who had not yet joined the union. The union was also sensitive to the grief of their ethnic brothers, being careful to appoint Slavic-speaking members to coordinate funeral and burial arrangements with the Slavic societies.[135] There were some cases, however, where ethnic ties were stronger than union brotherhood. During the strike of 1918–19 at Coal Creek, a representative from the Italian community approached the mine management, stating that they were anxious to go back to work and that they could persuade the other foreign nationalities to join them.[136]

The unions provided other assistance to the miners, particularly in times of grief and disaster. The union regularly paid a standard $100 toward the funeral expenses of one of their members, and often contributed more.[137] In the aftermath of major explosions, the unions took up collections for each other's locals. A motion from the Coleman Local 2633 stated:

> ... every member of the union give $1.00 to help those who have lost their support through the explosion at Mishell [sic] and that [certain] members of the union go around the citizens of Coleman to solicit money from them for the same.[138]

The union also assisted individual members when they were in need. When Edgar Ash lost his right leg in a mining accident in 1908, the International Coal and Coke Company shared the expenses with the UMWA to send Ash to Minneapolis for an artificial limb.[139] Since Ash could no longer mine coal, the union also may have had a hand in ensuring that he retain a job as the lampman, a position he held for the next forty years. Abraham Dodd, an old miner with no family in the Crowsnest Pass, died of exposure in 1926. The union immediately stepped in and paid out $213.40 for his burial.[140] In 1905, the Coleman Local made a motion that their membership should fund, build, and maintain a hospital. The union would run the hospital themselves, paying the doctor and nurses out of a fee levied on all members. The hospital was to be restricted for the use of union members and their families, unless non-members could pay the fee of $20 per week.

There were times when the union appeared unsure of the amount to be given out in assistance. When Henry and Harry Grewcutt died in 1907, the local kept tabling the motion to give aid to the widow. They then decided to do the same for Mrs. Grewcutt "as was done for others that is an average day's pay." At subsequent meetings this motion was rescinded, then reinstated, and eventually followed by a general motion to pay out "$500.00 to all widows whose husbands were killed in and around the mines." A year later, this motion was also rescinded.[141] There is no indication of how much Mrs. Grewcutt eventually received, although it does appear she was in desperate need of the money. She was in poor health and Henry had apparently not been well himself before his death, which may have impaired his earning power. One of the last comments that can be found for the Grewcutts does not bode well for their fate: "His widow and family are practically left to charity."[142]

The assistance provided by the unions, the fraternal organizations and the ethnic mutual aid societies was of particular importance in the days before compensation, and even afterward, since compensation could prove elusive to receive or qualify for.[143] One British Columbia mine manager's view in 1905 speaks for itself: "[The] Workman's Compensation for Injuries Act is the most damnable piece of legislation that any respectable people have to deal with in any country."[144] Alberta lagged behind British Columbia in proclaiming a Compensation Act, but the attitude among managers was still the same as evidenced by the testimony given to the Alberta Coal Commission in 1907: "If a man thinks he can get a claim he will not be so careful."[145] Other men speaking at the commission were of the opinion that it was incumbent on all miners to insure themselves against accidents, sickness, and death. Yet insurance agents were asking $25 per year for accident coverage, an expense few men could afford given the other claims on their paycheques.[146]

When disaster struck a family and the wage earner died in the mines, the families were frequently left with few resources. William Watkins died leaving his widow Annah with a total of $61.08; when

Herbert Ash was killed he left behind less than $100 to support a wife and four children. The Falip brothers, Jules and Henri, both died in the 1917 Coal Creek explosion, leaving each a wife and child. One year later, Henri's widow Marie also died, leaving Jules' widow Erma to look after not only her own family, but Marie's daughter Alice, aged two. Erma received $2 per month from the estate for the care of Alice, an amount that never varied between 1918 and 1937.[147] Some women left the country entirely and attempted to navigate the complex process of compensation from Italy or Poland or Austria. But many others stayed in the Crowsnest Pass and tried to make a living for themselves. Some women remarried: of the thirty-two Anglican women married between 1917 and 1930, four were widows.[148] Nellie Emerson petitioned the Compensation Board for sufficient monies to open a boarding house in Coleman, while other women took in washing to make ends meet.[149] Elizabeth Kidd lost first her son to the Blairmore mines and then her husband, but fell back on her own training to open the Kidd Maternity Home to support her remaining four children.[150]

Churches

The Crowsnest Pass churches also supplied community support in times of disaster. Both the Methodists and Presbyterians arrived early in Fernie in 1898; both soon established times for additional services in Michel, Morrissey, Coal Creek, Hosmer, and Corbin. The Anglican church first laid its cornerstone in 1900 in Fernie.[151] By October 1901, the *Fernie Free Press* listed services by the Methodists, Christ Church, Holy Family, Salvation Army, Knox Church, and the Baptists. The Methodists especially took an aggressive approach, advertising on the front page of the Fernie newspaper for years. The central location of Fernie and its role as a service centre to the entire mining area on the British Columbia side of the border probably gave some assistance to the various churches in maintaining steady congregations. However, while the churches in the Crowsnest Pass had the infrastructure in place to respond to tragedies, the role of each church appears plagued by problems in keeping personnel in the district, and by the community's sporadic involvement.

On the Alberta side of the border, the Anglican Church suffered an early setback at Frank, since the first minister had the unfortunate timing to arrive immediately after the Turtle Mountain slide: "The confusion resulting, and departure of many inhabitants made it appear inexpedient to remain."[152] The scattered nature of the towns on the Alberta side of the Pass made it difficult to keep a congregation focused on worship. The Anglican minister, Mr. Robinson, discovered this in 1906 while trying to visit Frank, Blairmore, Coleman, Lille, Hillcrest, and Bellevue in his round of duties. He received nothing but complaints from the people that they either did not see him as often as they would like or that the times he was forced to choose for his services did not accommodate their needs.[153] The mobility of the population combined with the debilitating effects of both work slowdowns and prolonged strike action to make it difficult for the Anglicans to keep a church together. In 1926, the Anglican church warden wrote that "about six people are keeping this church going ... there must be 50–60 more vacant houses than two years ago."[154] However, the service registers show an average of between twenty and forty people attending church, although this ranges from a low of two to a high of 125. The higher attendance figures usually coincided with children's services, or special ceremonies such as the Harvest Service.[155]

The Catholic churches on the Alberta side of the border had a far more difficult time becoming established and in keeping their membership.[156] The parishes fought against rapid turnover of their priests from ill health or disheartenment. Four priests served the parish of St. Anne's Roman Catholic Church in Blairmore over the space of three years from 1911 to 1914:

> There can be no doubt this work was difficult and discouraging for all of the four retired because of ill health. Poverty, indifference, even ill will on the part of some, the ugliness of the towns and the grueling means would have broken the strongest spirit.[157]

One of the main problems was the ethnic component of the towns. Father deWilde, who had charge of Frank, Lille, Blairmore, Hillcrest, and Bellevue, remembered:

> There was a pretty good number of Catholics [in Frank in 1908], principally of Irish and German descent ... there were also a few Slavonians, fewer maybe than elsewhere ... besides them many others remained but never set their feet in church ... I used to go around all the houses where I knew there were Catholics but I soon became aware that such a thing was absolutely useless. Lille could have at that time a population of 500 to 800 mostly Italians but very few Catholics.... Bellevue included Italians and chiefly Slavonians. As for Hillcrest a great part of the population was composed of French who unfortunately were very much opposed to religion.[158]

Father Beaton commented in 1917 that although "the Slovaks, Hungarians, Poles and Irish Catholics were as a rule very good, the Italians, French, Belgians and Bohemians ... acted as they do elsewhere."[159]

The priests worked exceptionally hard to reach their parishioners and shepherd the delinquent back into the church. Father Cosman arrived in the Crowsnest Pass in 1917 and immediately did a house-to-house canvas of the entire parish. He managed to round up three members who had not been to confession in over twenty years, eleven who had not attended in over fifteen years and twenty-five more who had been absent from the fold for at least ten years.[160] Five years later, however, Father Kreciszewski was still complaining that the people were not attending to their spiritual duties and prayed to God to "move their slothful hearts."[161] Other denominations also noticed the fall-off in attendance on Sundays. The Presbyterians launched a religious revival in 1909 in the Kootenay region that drew people back to the church in droves for a time.[162] Perhaps some of the early indifference to attending any service of any denomination came from the fact that much of the population in the Crowsnest Pass continued to be single men for some time. Marriages, baptisms, and confirmations of children were rituals linked to the church through families, and were not the province of itinerant workers.

There was also a continual pressing need to find men who could communicate with the different nations represented in the Pass. The steady stream of dismal stories that the clergy sent back to the Catholic Bishop concerning the "terrible, terrible" attendance at services are more than adequately balanced by the obvious need on the part of the townspeople, who pleaded almost constantly for a priest to hear their confessions in their own language. The Polish and Slavonic populations in particular were vocal in their need for a priest who did not require the use of a dictionary for them to do their duty to God. John Bielec, who would later die in the mines from lack of first aid, sent an especially heartfelt letter to the Bishop in 1916:

> ... we are all one faith, Roman Chatolic [sic] priest Bishop ... we are pleading you priest Bishop if you could send us a Polish priest. There are 47 Polish families [in Coleman] and men unmarried we don't know, Slavs 20 families, English 7 or more we don't know. This year very few Polish people went to confession because the priest was using a dictionair [sic] because he couldn't confess in Polish. All the Polich people are Chatholics [sic] and today many from the Chatholic faith are taken away because he doesn't hear a word of God although he would go to the Church because he little understands English tongue.... [163]

The Polish population sent an unsuccessful petition with close to fifty names to the Bishop in 1923 requesting the return of Father Joseph Kreciszewski, the only priest they had who could speak their language.[164] Father DeWilde of Coleman was an exception, as he spoke French, German, English, Italian, and Flemish. DeWilde retired due to exhaustion after three years in the Pass.[165] On many occasions, the Alberta Diocese took advantage of the more settled Catholic Church on the British Columbia side of the border. In 1906 Father Messner arrived from Fernie to hear the confessions of the Slavs and Polish; Bishop McNally stated in 1917 that Father Sworski could be loaned from Fernie as often as needed; Father Aloysius came through the Pass in 1926 to hear the Italian confessions.[166]

The Catholic Church faced the same problems as other denominations in attracting parishioners due to the cyclical nature of mining work. This could lead to poverty among the parishioners, which certainly "did not promote a contented and ambitious community." Many of the miners came from Eastern Europe as well, where anticlericalism was rife. Referring to this problem, the priest from St. Cyril's Roman Catholic Church in Bellevue stated bluntly in 1927: "We have 750 European Catholics in this parish, only about 50 of them attend Mass."[167] In 1915, Reverend Beaton of the Bellevue parish wrote to the Bishop: "The Oblates found [them] very indifferent towards the practice of the faith and towards the support of the priest and the Church."[168]

Despite the problems of language, anticlericalism or general antipathy, the churches in the Pass undoubtedly offered comfort to families in need. As a network of support during times of great stress and disaster, the churches would have responded with everything at their disposal, both physical and spiritual: "The presence of the church, unaccepted by many, was appreciated most in the midst of disaster."[169] The facts that the majority of funeral services are invariably mentioned as "taking place from the church," and that the clergy are present in photographs when supporting the bereaved bears testimony to the work of the churches in support of the larger community. Although the ministers and priests may have had difficulty themselves in the face of overwhelming disaster, when the gaping holes left in the community would have "disheartened the bravest of priests," the church provided much-needed solace to individuals and their families when called upon to do so. To those of faith, the response of the church and their interaction in the lives of families who had suffered a loss would unquestionably be of great comfort.

The Media

Major explosions upset the accepted continuity of life. In the natural order of things, people died individually after a certain span of time, allowing for the family to grieve and bury the dead with some privacy. Mining disasters disturbed this expected order, resulting in "unnatural collective deaths."[170] The usual privacy that attended an individual's death was no longer a factor. "Collective death means that the normal screens that hide death in everyday life are taken away and everything about it can be seen."[171] Nowhere was the public nature of death in mining disasters more apparent than in the media.

The *Calgary Daily Herald*'s accounts of the 1910 and 1914 explosions are expansive in their descriptions of survivor's grief, of women wailing and weeping, standing with red eyes, or collapsing on the ground in front of the coffins of their dead. However, these accounts of the physical manifestations of grief, although present, were far more restrained in the community papers of Bellevue, Blairmore, and Coleman. There could be several reasons for this. The Calgary papers had a different audience, and perhaps played to the expectations of that audience. The community papers, given their limited and more intimate readership, may have tried to downplay the more obvious aspects of the area's loss, given the familiarity and knowledge that most of their readers had regarding their neighbours' grief. Or, this muted local coverage may have had something to do with the more general acceptance of the nature of the work in a mining town and the sense of fatalism fostered by the miners.

The media also played another role in major disasters. No disaster discourse seemed complete without the inclusion of hero stories, accounts of daring rescues, individual bravery, and sacrifice. Following the 1910 Bellevue mine disaster, headlines proclaimed that the miners "Died Like Heroes."[172] In the midst of mass death, these stories served a purpose. They reaffirmed the possibility of miracles; if one man could escape by daring deeds, then perhaps others could as well. Hero stories also gave solace to those whose fathers, brothers, or sons had died. Who knew but that the dead may have also exhibited the reported actions of living heroes, their stories remaining untold by their last silence? Projecting the actions of others onto the dead did no harm, and it may have given some consolation to those in mourning.

Heroic figures who lived could provide eyewitness accounts that might have given some support to wives, mothers, and sisters trying to imagine the last moments of their loved ones. During the 1926 McGillivray mine explosion, despite the presence of gas and fire in the mine, Hugh Dunlop took the time to chalk directions on the mine timbers as he led other miners to safety.[173] I.A. James stayed in the mine for a considerable period at risk to his own life in 1926 in order to telephone instructions to the surface on the condition of the mine and the location of the men.[174] The popular media celebrated the heroic nature of all miners. In an interview after the 1910 Bellevue explosion, a member of the rescue team stated: "You asked me if I can tell you of any particular man that was extra brave ... I think it would be just as well to give a list of every man that was there."[175]

Songs underlined the day-to-day heroism of miners. "A Miner's Life" stated that "ev'ry day his life's in danger/Still he ventures being brave,"[176] while "The Man with a Torch in His Cap" compares miners to soldiers:

> There are men who are heroes on battlefields,
> Who for freedom give their all.
> There are men whose lives have been scarred by war
> When they heard their countries call.
> And many a miner has laid his head
> In death on the coals black lap.
> So don't forget, he's a hero too,
> The man with a torch in his cap.[177]

Heroes who died also served a purpose for the community. They provided an outlet for grief that was perhaps not as personal or as painful for those in the community who had suffered a more immediate loss. Fred Alderson became the best known of the Crowsnest Pass mining heroes. Alderson joined the Hosmer Mine Rescue team just weeks before the 1910 Bellevue mine explosion. Since the members of the Hosmer team were the only well-equipped and trained group in the vicinity, they were immediately transported to Bellevue to assist in the rescue attempts. While attempting to carry a Draeger device to the men trapped inside the mine, the fumes overcame Alderson and he died.[178] Some confusion was evident in the newspapers regarding whether he was wearing his own Draeger device at the time, or whether the device had malfunctioned. However, in a later report on the disaster, James Ashworth, then the manager of the Crows Nest Pass Coal Company, stated that: "Alderson is said to have lost his life through some leak – the fact is that he was not wearing any apparatus, and therefore could not suffer from any leakage."[179]

Far more important than the exact details of Alderson's death was the fact that he had died in an attempt to rescue trapped miners, particularly since reports described him giving up his own Draeger device in a vain attempt to save another man. His death caught the imagination of the community and the press; the *Calgary Daily Herald* especially lauded Alderson: "Among all the heroes of the night and day of toiling rescue work the conscious and shining light is a man named Alderson."[180] R.W. Coulthard, the General Manager of the Bellevue mine, went so far as to insert a "card of thanks" in the newspaper, "above all ... to make mention of the gallantry and fearlessness of Fred Alderson, of Hosmer, who laid down his life in that most noble of all causes, the saving of human life."[181] Alderson's home county of Durham, England even mentioned his passing, reprinting the story of his death from the *Lethbridge Daily Herald*.[182]

Relief Funds

Alderson's death also provided a catalyst to another community disaster response – the growth of the relief fund. While relief funds were usually started after a disaster in order to benefit all victims, in Alderson's case, the fund was specifically for his wife and family.[183] The contributions were overwhelming as the community responded to the moving accounts of his death. Three months after the disaster, the fund had reached a total of $4,795.91.[184] Among the largest contributors were the Hosmer mine employees ($699.25) and the West Canadian Collieries that managed the Bellevue mine ($500). Arthur L. Sifton, then Minister of Public Works, also recommended to the Lieutenant Governor of Alberta that $500 be given to the Alderson Relief Fund "in recognition of [his] services."[185] On a sad note, one of Alderson's daughters later disclosed the fact that "all the money sent by the people of Canada was lost, as the new railway across Canada folded up."[186]

Relief funds provided an outlet for individuals and community to show their shared grief and empathy to those bereaved. The funds also helped people in the larger community to demonstrate their own feelings. Subscribers to the Alderson Relief Fund included individuals and companies from Calgary, Granite Creek, Carbondale, Kingston Ontario, Kaslo, Canmore, Nelson, Taber, Lethbridge, and Vancouver.[187] Disasters caught the imagination and many felt the general impulse to help in some way. The relief fund gathered after the 1902 Coal Creek explosion eventually reached upward of $40,000. Eight years later, an accounting of the use of the fund showed that payments had been made to the widows and orphans in the amount of $31,977.26. In a twist on local heroes, the administrators of this fund were lauded with "great praise" for the time and effort they had put in over the years overseeing the dispersal of the monies.[188]

Relief funds could also be viewed cynically. In the days before compensation claims, and even afterward, relief funds provided an easy response for the mining companies and the government to endorse aid for bereaved families without assuming responsibility.[189] After the 1902 Coal Creek explosion, the mine managers released a statement saying that they would pay all funeral expenses and "relieve against immediate want and suffering."[190] Three days later, the company amended this statement, worried about "the very harmful effect that circulation of such a report would have upon the contributions to the relief fund."[191] The management did not want to "impede the flow of contributions" from others and to its credit the company did not withdraw its initial offer of help.[192] The Crows Nest Pass Coal Company contributed $10,000 to the fund.[193]

Mourning

If grief is the first emotional reaction to death, then mourning is "the culturally patterned way in which bereavement is expressed."[194] Mourning begins with the funeral; in mining towns, funerals were significant rituals.[195] All miners faced the same dangers, and could ultimately face the same fate. Funerals therefore marked a transition not only of a single miner or of several killed in an explosion, but could also foreshadow the death of any number of the miners who marched in solemn procession behind the funeral wagon. In general, funeral processions allowed a community to show respect and to provide dignity and a sense of importance to the passing of their own.[196] In the Crowsnest Pass, funeral processions were also a tangible symbol of the grief of the entire community, since danger and death were central to all their lives. The community marched in the processions not so much for the sake of the dead, but as a public act of mourning for the sake of the living.[197]

Music attended all Crowsnest Pass funerals. The *Coleman Journal* announced in 1926 that "a good band is one of the most useful things a town can possess."[198] The bands played at all social events, dances, benefits, and weddings. The Pass had a wealth of musical talent, so much so that sometimes good-natured rivalries sprang up between the bands. The Blairmore and Bellevue bands held a "play-off" in 1914 to see which one of them warranted the most praise.[199] The bands were of vital importance to funerals. For many cultures, music at a funeral was a given, providing "one of the most profoundest means of expressing emotion."[200] Bands were common at Italian funerals,[201] while Polish and Ukrainian funerals were matters of great moment, with music and song evident during the wake, the funeral and the funeral feast afterward.[202] In the Crowsnest Pass, the band would take their position at the head of the funeral procession to lead the mourners to the cemetery (see Photo 10). For most Crowsnest Pass funerals, the band played the Dead March going to the cemetery, followed by "The Girl I Left Behind Me" on the way back.[203]

Funeral processions usually formed with the clergy and/or members of societies or fraternal orders in the lead, followed by the flower carriage, honorary pallbearers, active pallbearers, the hearse, immediate family, relatives, and friends.[204] Crowsnest Pass funerals were frequently large events with the procession stretching for blocks (see Photo 11). In the case of the larger disasters, practically the entire community took part in the funeral processions. In the aftermath of the 1902 Coal Creek explosion, the entire town of Fernie was shut down for a day while a series of thirty funeral processions were held.[205] Following the 1910 Bellevue mine explosion "about 2,000 people wended their way from the cemetery."[206]

Funerals for individuals killed in the mines were also large events. When William Anthony was killed in 1904, the miners' union chartered a special train from Michel bearing the corpse and 140 friends and relatives to Fernie, where the funeral took place. Anthony was a member not only of the union but also of the Knights of Pythias and the Oddfellows. Each organization marched in the procession and performed their ceremonies at the graveside.[207] When Emile Genjanbre died in a coalfall, over three hundred attended his funeral. Among the attendees were many members of both the Blairmore and Coleman Order of Oddfellows.[208] One of the most important ways in which fraternal societies and unions assisted in the funerals of their members was the stipulation in many of their charters that members must attend the funeral of one of their own or pay a fine. While this appears to be a rather magisterial command, or simply a way for the society to collect dues from its members, this custom served an important function. By marching in the funeral procession, the members of the fraternal orders and unions showed not only their respect for the dead, but also provided tangible evidence of their support for the family and the community.

Women played an important role in burying and mourning the dead. One wife in Fernie remarked that in the fourteen years she had lived in the area, she had attended all the funerals, "following [the] solemn funeral corteges up onto that hill."[209] The Rebekah Order also tended the graves in the cemeteries on a regular basis.[210] Regardless of how well known or how little, the dead had been part of a community, and the community had not forgotten them.

Funerals are the "shock absorbers of death,"[211] the ritual where mourning and acceptance of loss began. Besides their religious or ritualistic nature, funerals are important social events that can assist in the mourning process.[212] The social nature of funerals is evident from a blistering editorial in the *Pincher Creek Echo*. The author, angry over the apparent lack of deference shown while the Dead March was played, sternly admonished the crowd to improve their behaviour by exhibiting respect for the theme in the future. A week later, the paper underscored its point, with an obvious mention of the crowd standing quietly at a funeral during the playing of the march.[213]

Marking Death

One of the final responses of individuals and communities to death and disaster are the choices that the living make to commemorate their dead. A study of grave markers, monuments, and memorials can often yield revealing clues to the culture and value of individuals and communities.[214] Grave markers and monuments can serve to display individuality of both the deceased and the ones left behind.[215] In the case of the largest mining disasters, the numbers of dead sometimes mitigated against the response of a personal grave marker. Following the 1902 Coal Creek, a mass gravesite was chosen for the dead on a bench of land halfway up the cemetery. The graves were dug with twenty-five in a row to accommodate the 128 coffins. After the 1914 Hillcrest mine explosions, the majority of the 189 dead were

10　Funeral of Frank Malec, 1923. Crowsnest Museum, CM-BL-13-14.

POHŘEB FRANT. MALCE Z FRAN
ALTA, KTERÝ STRATIL SVŮJ ŽIVOT
UHELNÉM DOLU DNE 7th ČERVENC

11 Funeral Procession, Main Street Blairmore, c. 1924–30.
Crowsnest Museum, CM-BL-13-02.

also buried in two long rows, although individual stones are also located in the area (see Photos 12 and 13).

However, in both instances, some personal choices in regards to individual grave markers are evident. At the edge of the mass grave in Fernie, where the victims of the Coal Creek explosion were laid, stands a large marble monument of a tree carved into a cross with the truncated limbs that symbolized a life cut short. The inscription reads:

> John Hovan
> Aged 16 years
> Came to his Death by Explosion, May 22 [Photo 14]

James Mitchell, also killed in the 1902 Coal Creek explosion, had a far more elaborate grave marker raised in his honour. The marker is massive, ringed by iron railings and surmounted by a white marble angel (see Photo 15). Mitchell did, however, have to wait thirty years for the marker, which he shares with his wife. Mary Ann Mitchell

12 Mass Graves at Hillcrest Cemetery. Photo by Karen Buckley.

13. Mass Graves at Fernie Cemetery. Photo by Karen Buckley.

DISASTER 119

14 Grave Marker of John Hovan, Fernie Cemetery. Photo by Karen Buckley.

15 Grave Marker of James Mitchell, Fernie Cemetery. Photo by Karen Buckley.

16 Grave Marker of Thomas Bardsley, Hillcrest Cemetery.
Photo by Karen Buckley

122   DANGER, DEATH AND DISASTER

17 Grave Marker of Henry Plasman, Coleman Union Cemetery. Photo by Karen Buckley.

had the whole piece shipped over at great expense from England to commemorate the death of her husband.[216] In contrast, the Hillcrest cemetery has a much simpler stone marking the place where Thomas Bardsley lies. Bardsley died in the Hillcrest mine explosion with his pick still clutched in his hand. The only evidence of the force of the blast that killed him was that his collar had been torn off. Bardsley's grave marker has a spare beauty to it, a small oblong stone with a spray of raised lilies and a banner with his name (see Photo 16).

Bardsley's natural spray of lilies is evocative of grief even without words. Far more common on miner's grave markers are the words of family who wanted some note recorded regarding the manner of their dying. The families of Henry Plasman and Joseph Homenick chose the shorter epitaphs of "Killed in Mine" or "Killed in Coleman Mine" respectively (see Photos 17 and 18). The families of Henry Grewcutt and Samuel Edwin decided to elaborate this epitaph somewhat by engraving "Accidentally Killed" on the stones (see Photos 19 and 20), while George Lothian's wife provided even more detail. Lothian's stone reads:

> Killed by Gas, Nov. 17 1928
> In International Mine. [Photo 21]

Perhaps the overwhelming emotions at the 1914 Hillcrest mine disaster led to the choice of words on the grave markers of the men who had died there. The families of Robert Smith and Charles Coats responded to their loss by carving: "Lost in Hillcrest Explosion" (see Photos 22 and 23), while the single stone commemorating the death of four members of the Murray family also reads "Lost in Disaster." The evocative choice of words on these grave markers leaves behind a haunting image of these men's fate.

Grave markers could be either handmade or mass produced. After the Coal Creek mine explosion of 1917 killed thirty-four men evidence suggests that a single shipment of grave markers was commissioned. The majority of these stones are grey marble obelisks, all engraved identically with the same words in the same order. Jules Falip's marker provides an example (see Photo 24):

> In loving memory
> of

18 Grave Marker of Joseph Homenick, Coleman Roman Catholic Cemetery. Photo by Karen Buckley.

19 Grave Marker of Henry and Harry Grewcutt, Coleman Union Cemetery
Photo by Karen Buckley.

20 Grave Marker of Samuel Edwin, Hillcrest Cemetery. Photo by Karen Buckley.

21 Grave Marker of George Lothian, Coleman Union Cemetery. Photo by Karen Buckley.

22 Grave Marker of Robert Smith, Hillcrest Cemetery. Photo by Karen Buckley.

23 Grave Marker of Charles Coats, Hillcrest Cemetery. Photo by Karen Buckley.

Jules Falip
Killed
Coal Creek
April 5 1917
aged 36 years 6 mos.

Identical markers are found nearby for Henry Falip, Joseph Campbell, Willie Sherwood, John Monks, and William Puckey among others. In contrast, William Clarke, who was also killed in the 1917 explosion, is buried beneath a hand-lettered piece of white slate, surrounded with roughly carved wood:

In Memory of
William Clarke
Who was Killed in the
Explosion at Coal Creek
on the Fifth Day of
April 5th [sic] 1917.

Clarke's son is also commemorated on this tombstone (see Photo 25).

Heroes are duly commemorated as such on their stones. Fred Alderson's sister chose to mirror the words of praise for her brother that she read in all the newspapers. The *Bellevue Times* had headlined an article on Alderson with the quote: "Greater love hath no man than that a man lay down his life for his friends," going on to state that "a nobler deed could never be done."[217] Alderson's sister engraved the same quote in full on his monument. Alderson's large and ornate red granite monument stands alone in an otherwise almost markerless Hosmer graveyard (see Photo 26). Similarly, Evan Evans, another miner of heroic stature who died following the 1915 Coal Creek explosion, was also lauded for his efforts on his gravestone. The words engraved on his large marker are:

In Memory of
Evan Evans
Inspector of Mines
Late of Glamorganshire, Wales
Died in the Execution of
his duty at Coal Creek

January 2 1915
Aged 42 Years.

Evans, the natural son of a domestic servant from Wales, died without a will or any mention of relatives, and with a large sum of money to his name. When he was buried, the government of British Columbia paid all funeral expenses, "owing to the great esteem in which the late Evan Evans was held throughout this district."[218]

In the examples given above, these grave markers demonstrate the visible responses of families and communities bereaved in the Crowsnest Pass. The examples are significant because they show how many of the grave markers were fashioned to note the manner of the men's dying, not just the fact that they had died. Some record apparently needed to be left behind, either for the families' sake, or for the community's, so that the details of these men's deaths would not be forgotten. Other graves are not so well marked. Although the gravesites of both Albert Nycoma and Fred Iwaniczuk are indicated in the Hillcrest grave directory, neither of these plots could be located in the cemetery. They are just two of those who died in the mines who were buried but not commemorated by family or community in any way. Perhaps a simple wooden cross, long since decayed, marked the grave. For others who died in major explosions, the mute testimony of the communal graves perhaps says all that need be said of their passing.

Interestingly, none of the grave markers of the men killed in the 1926 McGillivray mine explosion whose stones could be located have anything carved on them to mark the manner of their death. The only notation on these markers is "Died." These men were not "Killed," "Killed Accidentally," or "Lost" as noted on other grave markers. The change in epitaph can perhaps be traced to other events at the time: both the 1914 Hillcrest mine and the 1917 Coal Creek explosions took place during World War I. Engraving "Killed" or "Lost" on a monument might reflect the standard vocabulary at the time, and not refer to the emotional response of the families. However, there are exceptions: Earl Eckmeier's grave marker reads "Killed July 10 1923," while Peter Blane Anderson's 1922 epitaph notes that he was "Accidentally Killed."

24 Grave Marker of Jules Falip, Fernie Cemetery. Photo by Karen Buckley.

25 Grave Marker of William Clarke, Fernie Cemetery.
Photo by Karen Buckley.

26 Grave Marker of Fred Alderson, Hosmer Cemetery. Photo by Karen Buckley.

Although a thorough comparison has yet to be done on miners' grave markers in general, the notations found on Crowsnest Pass graves do not appear unique. The cemeteries of Nanaimo and Cumberland, major mining centres on the coast of British Columbia, are full of similar types of stones and epitaphs, many of them stating, "Killed in explosion," "Killed in mine explosion," or "Killed in No. 5 mine." The Nova Scotia cemeteries also contain markers engraved with the details of mining deaths, including some relating to the heroic nature of miners who died trying to rescue others in the mines.[219] Epitaphs are also noted in popular mining songs. The chorus of "Only a Miner Killed in the Breast," states: "Read on his grave 'neath the whispering pines/ *Our Joe, aged just 20, was killed in the mines.*"[220]

What could perhaps be charted in greater detail are the types of monuments and memorials created in mining areas. Monuments denote the large, communal edifices to a specific mining disaster, while memorials include songs, poems, "In Memoriam" notices in the newspapers or anniversary services. Memorial songs are numerous and include "The Unknown Miners Grave," Miner's Epitaph," and the "Miners Memorial Hymn."[221] Songs specific to the Crowsnest Pass area include "The Ballad of the Frank Slide"[222] and the popular folksong by James Keelaghan on the "Hillcrest Mine."[223] More recently, Monica Fields of the Frank Slide Interpretive Centre composed the song "Hear That Whistle Blow" in 1985 to commemorate the 1914 Hillcrest mine disaster.[224]

Other types of memorials were also published in the Crowsnest Pass in the form of poems or notes inserted in the newspapers on the anniversary of men's deaths. While the men who died in the 1926 McGillivray disaster did not have the manner of their dying engraved individually on their stones, they were among the few who warranted this other type of memorial. Poems to the McGillivray victims appeared in print on 10 February 1927 and a year after their deaths on 24 November 1927. Other individual mining deaths were also remembered in print. Joseph Burns died in the Coleman mine on 21 December 1925. One year later his family inserted a poem commemorating his loss. Frank and Joe Smith, father and son, were killed in the Coal Creek explosion of 1917. In memoriams to them were inserted in the 1921 and 1925 papers by members of their family.

Perhaps significantly, these types of public notices were rare. One author has suggested that "In Memoriam" notices and poems were inserted to "help the bereaved to maintain and re-establish a sense of social and biological continuity in their life-histories."[225] Although there were scattered print memorials commemorating other deaths such as mothers or infants in the early 1920s, this general practice of commemorating the dead through print apparently only gained significance in the Crowsnest Pass in the later 1920s. Again, perhaps the pragmatism of mining families assisted in accepting the loss of their dead. Family commemorations may have been private affairs with no need felt to mark the anniversary of the death in a more public way.

The dead were publicly honoured in other ways. The fraternal orders frequently held memorial services for their dead such as those carried out by the Oddfellows, Eagles, and Rebekahs.[226] Other memorial services were also held by the community to commemorate mining disasters. The 1910 Bellevue mine explosion was commemorated in a one-year anniversary in December 1911.[227] Anniversary memorial ceremonies in 1916 and again in 1925 commemorated the 1914 Hillcrest mine disaster.[228] Public ceremonies to mark the anniversary of 1902 Coal Creek explosion were noted one year after the event; eight years later a major memorial service was held on the Crows Nest Pass Coal Company's lawn consisting of clergy, the mayor, the union, and several fraternal societies. A concert was held again in Fernie in 1913 to commemorate the dead.[229]

While these efforts appear sporadic when compared to other mining communities that held and still hold annual commemorative events,[230] the pattern is actually similar. The second and third generation descendants have since picked up the need to commemorate the community's dead. Since its inception in 1985, the Frank Slide Interpretative Centre has commemorated the Hillcrest disaster every year, holding larger special events every five years. Since the 1950s, numerous community groups have attempted to install a larger monument to commemorate the Hillcrest mine disaster. The Crowsnest Pass community achieved this in the year 2000 with an impressive monument at the entrance to the Hillcrest cemetery.[231] The monument commemorates not just the Hillcrest disaster, but lists all major mine explosions throughout Canada (see Photos 27 and 28).

Mining disasters left intact the essential infrastructure of the Crowsnest Pass communities. The primary group of victims was limited to the miners themselves – their families were only in emotional, not physical danger. In anticipation of mining accidents and the recognized need for mutual aid, some of the necessary coping mechanisms were already in place in the community, with the result that the Crowsnest Pass residents were perhaps marginally prepared for the after-effects of a major mine explosion. The miners themselves would have been sustained by the knowledge that in the event of a disaster and the possibility of being trapped or killed, the community would react in a predictable way:

> If the disruption imprisoned him [the miner] for a longer period, he had the assurance that the mine management knew his location in the mine. He knew from his knowledge of the miners' code that his fellow workers would toil ceaselessly until all the men (or bodies) were recovered.... A trapped miner was sure that his family ... could count upon support from kin, friends, and neighbors.[232]

This "code" that all miners depended on grew as a natural response to the dangers faced by all miners in their daily work lives. They also knew that the mine administration at least attempted to keep track of where they worked in the mine in order to be able to find them in the event of an accident. Mine rescue techniques were also a response that the miners depended on, even if they first depended on their own actions and reactions to keep them clear of danger or get them out in the event of a disaster. The final response that the miners knew they could count on was the infrastructure already in place in the community that would provide support, aid, and assistance to themselves or to their bereaved families.

27 Hillcrest Mine Disaster Memorial, Hillcrest Cemetery. Photo by Karen Buckley.

HILLCREST
JUNE 19, 1914

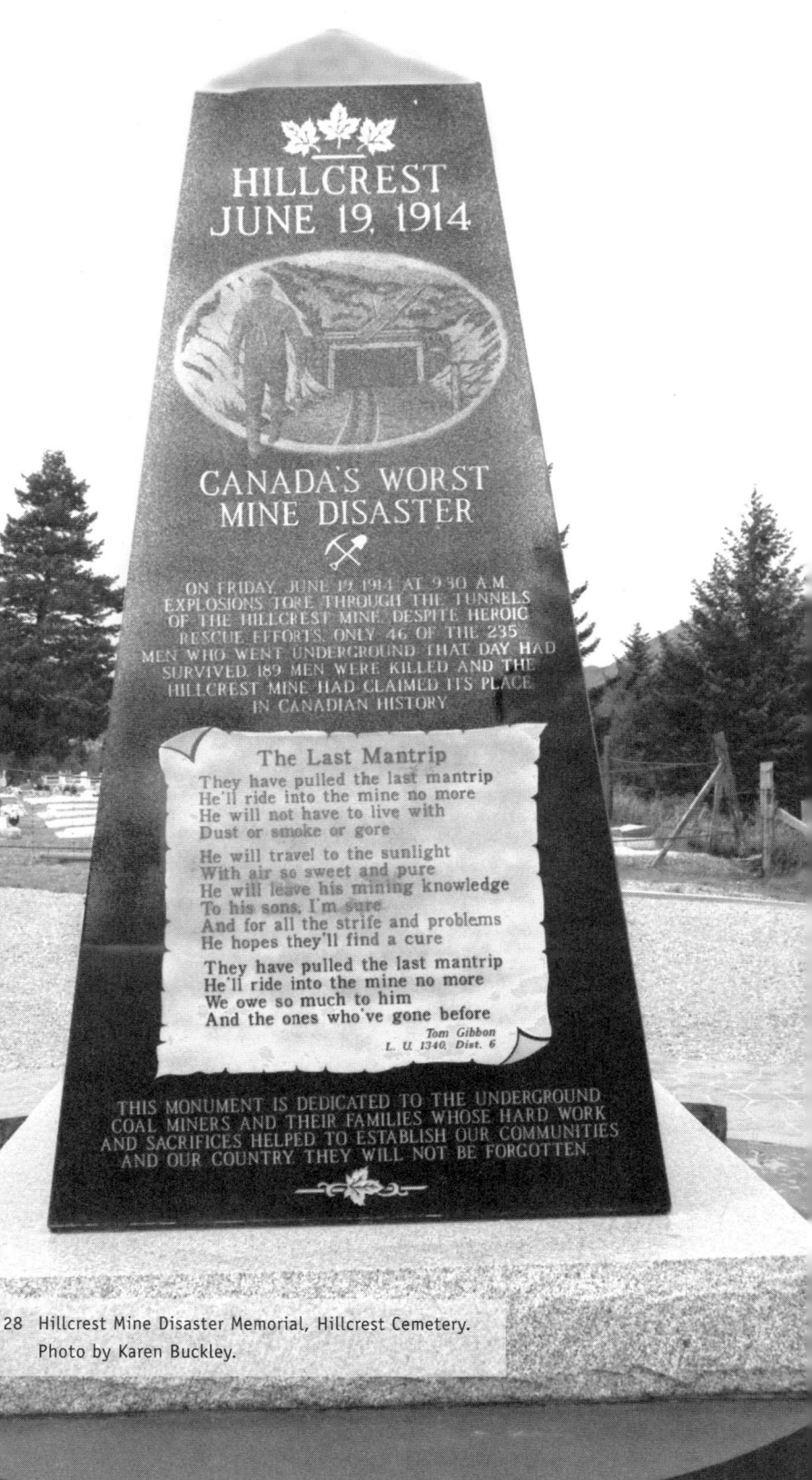

CANADA'S WORST MINE DISASTER

ON FRIDAY, JUNE 19, 1914 AT 9:30 A.M. EXPLOSIONS TORE THROUGH THE TUNNELS OF THE HILLCREST MINE. DESPITE HEROIC RESCUE EFFORTS, ONLY 46 OF THE 235 MEN WHO WENT UNDERGROUND THAT DAY HAD SURVIVED. 189 MEN WERE KILLED AND THE HILLCREST MINE HAD CLAIMED ITS PLACE IN CANADIAN HISTORY.

The Last Mantrip

They have pulled the last mantrip
He'll ride into the mine no more
He will not have to live with
Dust or smoke or gore

He will travel to the sunlight
With air so sweet and pure
He will leave his mining knowledge
To his sons, I'm sure
And for all the strife and problems
He hopes they'll find a cure

They have pulled the last mantrip
He'll ride into the mine no more
We owe so much to him
And the ones who've gone before

Tom Gibbon
L.U. 1340, Dist. 6

THIS MONUMENT IS DEDICATED TO THE UNDERGROUND COAL MINERS AND THEIR FAMILIES WHOSE HARD WORK AND SACRIFICES HELPED TO ESTABLISH OUR COMMUNITIES AND OUR COUNTRY. THEY WILL NOT BE FORGOTTEN.

28 Hillcrest Mine Disaster Memorial, Hillcrest Cemetery.
    Photo by Karen Buckley.

CONCLUSION

Henry and Harry Grewcutt were killed in the McGillivray mine in 1907, five years into a cycle of death and disaster that would affect the Crowsnest Pass until 1928. Henry was indicative of the type of men who were beginning to settle in the Pass, men who eventually married or brought their families, put down roots, had sons to follow them into the mine. Henry was already involved in two of the Pass's existing support networks, the church and the union. Had he lived, his son Harry would undoubtedly have continued to become as involved in the community as his father had done. The Grewcutts were spared by their deaths from the need to eventually join rescue crews themselves and venture into shattered and broken mines to find the bodies of their fellow miners. They would not have to witness the community trauma that followed the major loss of lives in 1914, 1916, 1917, and 1926. Nor would they have to march with other miners in the funeral procession for one of their own, or be called upon to read the burial service at the foot of an open grave.

However, the Grewcutts would also miss the chance to guide the union in its efforts to alleviate the dangers found in the mines and to take lessons in mine rescue and first aid. Harry would never have the opportunity to join a fraternity, to contribute to the comfort of individual miners through a shared drink after work, or to the community through relief funds or moral support. Henry would not be able to see his son married, or his grandchildren take part in the Sunday School he had helped to found. Neither would sit down to lunch with their fellow workers, or share a joke as they walked to the coalface, exchange songs and news as they splashed together in the wash house after their shifts, or pass on their knowledge, expertise, and pride to the next generation.

The mining industry in the Crowsnest Pass introduced men to an unnatural environment by taking them far underground away from the sunlight and open spaces. The mines were subject to disasters on a massive scale that killed over four hundred men in a twenty-six-year period and seriously injured as many more again. The combination of the physical environment and human factors made the mines of the Crowsnest Pass unique. The configuration of the coal seams, the volume of methane gas and coal dust, and the willingness on the part of many miners to take certain risks all served to accentuate an atmosphere of danger.

Men died from minor accidents such as a pick cut turned septic or a routine operation to fix a broken leg. They died in coalfalls, from haulage accidents, from major and minor explosions, from the effects of blackdamp and afterdamp, from the damage caused by bumps and blowouts. Some men died through their own carelessness, from inappropriate choices, and a disregard of safety precautions; but as many again died through sheer misfortune, seemingly at the whim of fate. Regardless if a man was a seasoned miner, or a greenhorn, it was not long before he learned first-hand of the danger lurking in the Crowsnest Pass mines. Whether they learned through word of mouth, familiarity from working in the mine, or listening to a more experienced partner, most miners and their families accepted, or at least tolerated, these conditions. Although some evidence suggests that a certain degree of mobility existed, the majority of miners chose to stay in the mines. If the miners either thought or knew that their prospects of finding other work was limited, they could develop an attitude of "learned helplessness," which in turn could lead to more frequent accidents.

Despite knowing the accident and fatality rate, men also stayed in the mines through a combination of economic need and family tradition as son followed father into the mine. Miners made passive choices to accept whatever the mine had in store for them by adopting a sense of fatalism. The miners tended to counter this by actively cultivating an awareness of independence and pride in their work. By instituting a sense of camaraderie with their workmates, miners implicitly gave themselves a support system which all knew they could rely on in the case of accidents or disasters. Miners were aware of, and frequently counted on, the fact that organized mine rescue crews could still save their lives in the case of a major disaster. When an explosion occurred, miners first relied on their own initiative and the assistance of their partners and friends to escape the dangers in the mines. However, when this first response failed, or when the explosion was so extensive as to negate their own efforts at rescue, miners knew that someone would eventually come looking for them. For many, the last chance they had to be pulled alive from certain death was through an organized mine rescue, despite the inherent problems.

When men died, families mourned their dead. Since death was so prevalent in the area, families could identify with each other's loss, either because of a similar death within their own family, or in the certain knowledge that death would visit them eventually. In the case of major disasters, the recovery of the dead and the identification of the bodies took place in front of the entire community. Death was not confined to one person in isolation, but a disaster that had to be shared by many in all its aspects. The community relied on the increasingly professional standards and duties of the undertakers, who were intimately involved in both single deaths and mass disasters. Whether a miner died alone in a coalfall, or as one of many in an explosion, the entire community participated in the funeral and the mourning process by marching in procession and lending support to the bereaved.

Support systems grew rapidly in the Crowsnest Pass to assist those dealing with grief and mourning. The creation of fraternal societies was a natural response to the knowledge of the precariousness of life in a mining community. The ethnicity of the Crowsnest Pass also provided the impetus for the development of other societies such as the Polish, Italian, Slavic, and Finnish Brotherhoods. These organizations lent a basic infrastructure to the community that could be relied on in terms of financial assistance, and more importantly, social and emotional support following the death of a member.

Despite the weekly tally of accidents, frequent deaths, and the interruption of grief and mourning, life went on. For some, life continued away from the Crowsnest Pass as wives and families moved back to Europe, Eastern Canada, or the United States. For others, life continued in the community as women remarried, or struggled to make ends meet as widows dependent on compensation claims and their own hard work to raise their families. Economic necessity sent most men back to the mines, even after a major disaster visited the town. Families buried their dead, some commemorated the lives of the lost, but most picked up the threads of their lives and created a new routine for themselves.

Many other mining communities suffered through one if not more of the same types of major loss of life and provide evidence of this continuity. Mining songs from across North America celebrate all aspects of the mining life, the everyday bravery of the miners, their fears, and their acceptance of the dangers. There is a commonality of responses to death and disaster, ranging from fatalism, a sense of independence and pride, a shared camaraderie, community and individual mourning patterns, to a growth in reliance on professional responses such as those evinced by undertakers and mine rescue personnel. The supporting infrastructure of churches and unions played a role in assisting individuals and the community, as did the fraternal brotherhoods and ethnic organizations.

The mining community in the Crowsnest Pass has long been studied due to its unique geographical location, its single industry, and the fact that since tragedy appeared endemic to the area, the communities grew to become a tight-knit group of people who assisted each other in these times of tragedy. The contrast to other mining communities across North America only underlines the commonality that existed between those communities and the Crowsnest Pass. Although unique in terms of the occurrence and constancy of danger, death and disaster, the mining communities of the Crowsnest Pass responded in a similar manner to like communities while dealing with grief and mourning. Men died and families grieved. Life continued.

Appendix A: Tables and Charts

Table 1: Disasters in the Crowsnest Pass

Date	Town/Mine	No. Killed	Reference/Source:
22 May 1902	Coal Creek	128	Frank Anderson, *Tragedies of the Crowsnest Pass* (Surrey: Heritage House Publishing, 1983), 9.
29 April 1903	Frank	2	J. William Kerr, *Frank Slide* (Priddis: Barker Publishing, 1990), 20. Although only two men died when the slide covered the mine, approximately 70 people died in the town and surrounding area.
15 October 1903	Morrissey	4	*British Columbia Annual Report for the Minister of Mines 1903.*
8 January 1904	Michel	7	*British Columbia Annual Report for the Minister of Mines 1904.*
18 November 1904	Morrissey	14	*British Columbia Annual Report for the Minister of Mines 1904*
3 April 1907	Coleman	3	Frank Anderson, *Tragedies of the Crowsnest Pass*, 8.
11 November 1907	Hosmer	3	*British Columbia Annual Report for the Minister of Mines 1907*
2 June 1908	Coal Creek	3	*British Columbia Annual Report for the Minister of Mines 1908*
22 July 1908	Coal Creek	4	*British Columbia Annual Report for the Minister of Mines 1908*
9 December 1910	Bellevue	31	*Annual Report of the Mines Branch of the Province of Alberta 1910*, 91.
19 June 1914	Hillcrest	189	Crowsnest Coal: Mining Disasters, http://www.crowsnest.bc.ca/coal09.html (26 March 2004).
8 August 1916	Michel	11	*British Columbia Annual Report for the Minister of Mines 1916*
5 April 1917	Coal Creek	34	*British Columbia Annual Report for the Minister of Mines 1917.*
19 September 1926	Hillcrest	2	George Rice, "Investigation into the Causes of the Explosion at Hillcrest mine," PAA 70.414.
23 November 1926	McGillivray Creek	10	*British Columbia Annual Report for the Minister of Mines 1926.*
30 August 1928	Coal Creek	6	*British Columbia Annual Report for the Minister of Mines 1928.*

Total Dead = 451

Note: The United States Bureau of Mines defines a mining disaster as one with five or more deaths. The other incidents are given here to note when more than one death occurred at the same time in one mine.

Chart 1

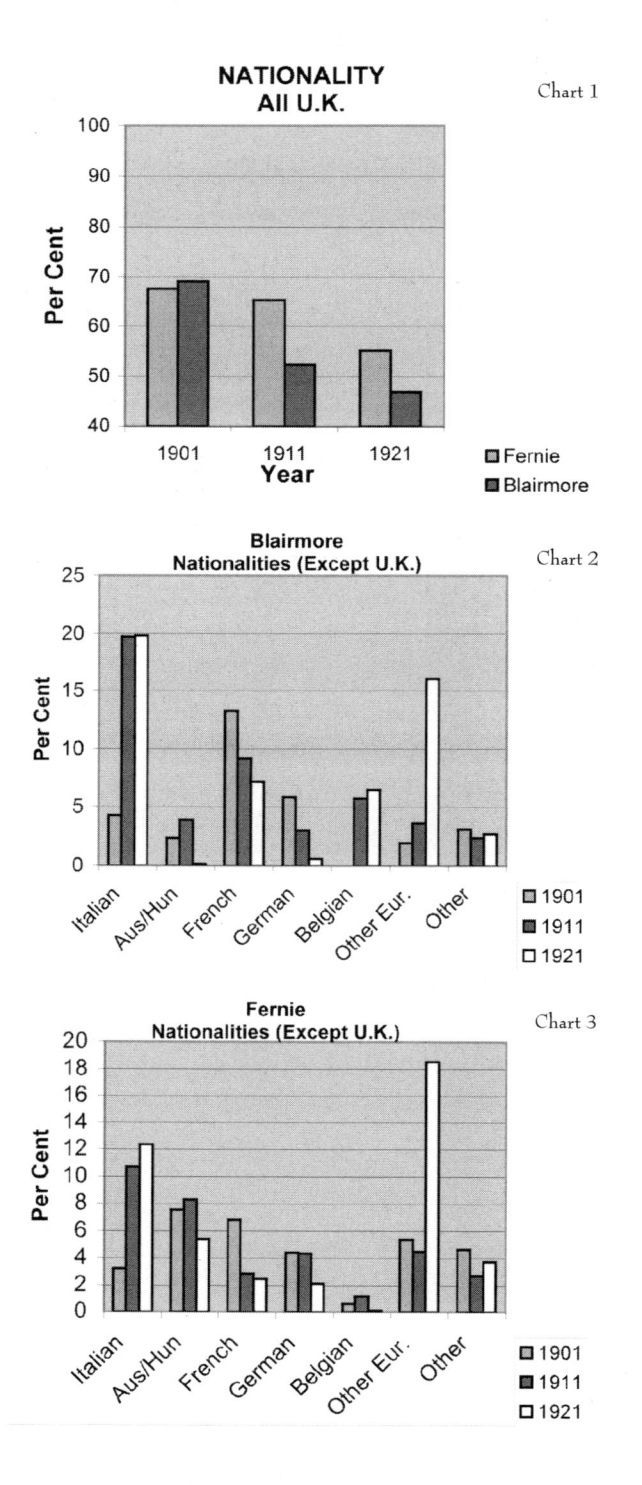

**NATIONALITY
All U.K.**

Fernie / Blairmore (1901, 1911, 1921)

Chart 2

**Blairmore
Nationalities (Except U.K.)**

1901 / 1911 / 1921

Chart 3

**Fernie
Nationalities (Except U.K.)**

1901 / 1911 / 1921

148   DANGER, DEATH AND DISASTER

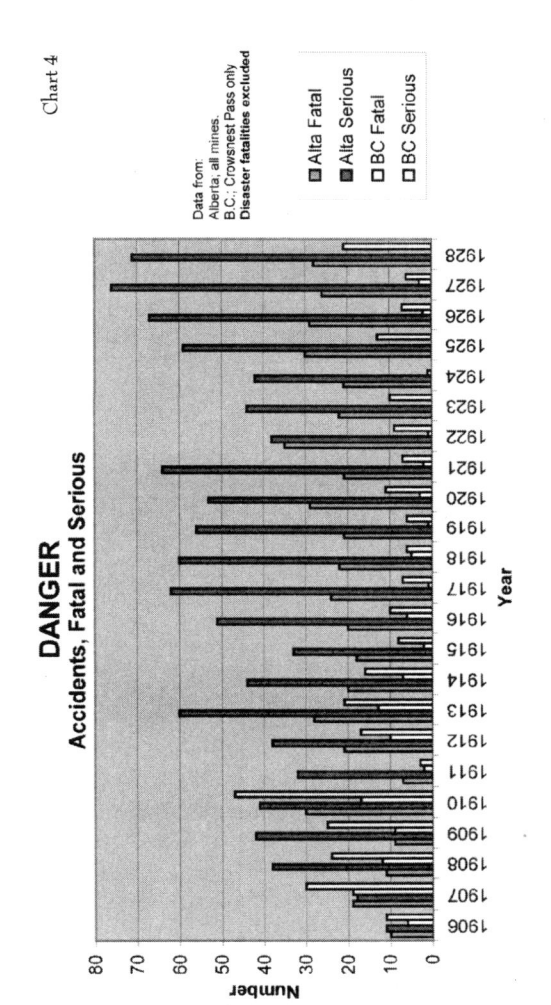

DANGER
Accidents, Fatal and Serious

Chart 4

Data from:
Alberta, all mines.
B.C.: Crowsnest Pass only
Disaster fatalities excluded

- Alta Fatal
- Alta Serious
- BC Fatal
- BC Serious

APPENDIX A   149

DISASTER
Fatalities in CNP Mine Disasters

Chart 5

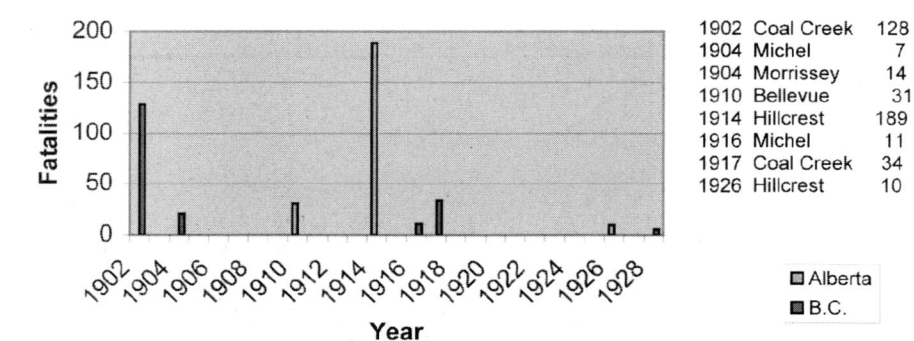

Year	Location	Fatalities
1902	Coal Creek	128
1904	Michel	7
1904	Morrissey	14
1910	Bellevue	31
1914	Hillcrest	189
1916	Michel	11
1917	Coal Creek	34
1926	Hillcrest	10

☐ Alberta
■ B.C.

**Fatal and Serious Accidents,
1904 - 1928
Data from all B.C. Mines**

Chart 6

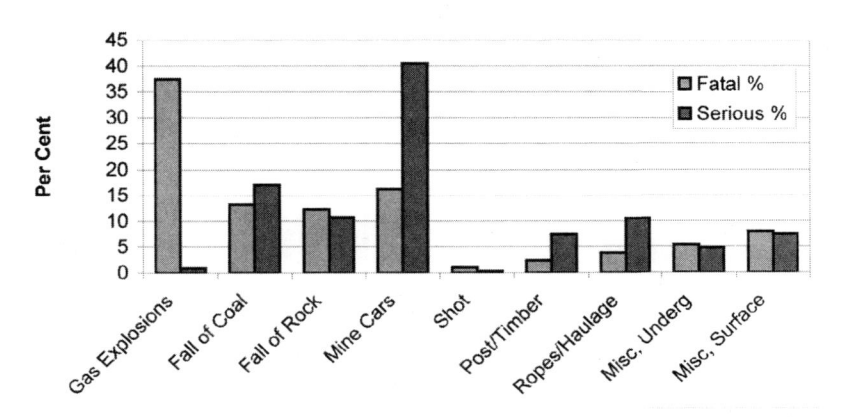

APPENDIX A   151

Appendix B: Mining Definitions

Blackdamp: A term generally applied to carbon dioxide or a mixture of carbon dioxide and nitrogen. An atmosphere depleted of oxygen.

Boss: Any member of the managerial ranks who is directly in charge of miners (such as shiftboss, fireboss etc.)

Brattice or Brattice Cloth: Fire-resistant fabric or plastic partition used in a mine passage to confine the air and force it into the working place.

Bumps: A violent dislocation of the mine workings which is attributed to severe stresses in the rock surrounding the workings.

Coalface or Face: The exposed area of a coal bed from which coal is being extracted.

Coalfall or Fall: A mass of roof rock or coal which has fallen in any part of a mine.

Firedamp: The combustible gas, methane, CH_4. Also, the explosive methane-air mixtures with between 5% and 15% methane. A combustible gas formed in mines by decomposition of coal.

Haulage: The horizontal transport of ore, coal, supplies and waste.

Pillars: An area of coal left to support the overlying strata in a mine; sometimes left permanently to support surface structures.

Robbing (or Drawing) Pillars: The systematic removal of the coal pillars between rooms or chambers to regulate the subsidence of the roof.

Room and Pillar Mining: A method of underground mining in which approximately half of the coal is left in place to support the roof of the active mining area. Large "pillars" are left while "rooms" of coal are extracted.

Safety Lamp: A lamp with steel wire gauze covering every opening from the inside to the outside so as to prevent the passage of flame should explosive gas be encountered

Timbering: The setting of timber supports in mine workings or shafts for protection against falls from roof, face or rib.

Trip: A train of mine cars.

White damp: Carbon monoxide, CO, a gas that may be present in the afterdamp of a gas or coal dust explosion, or in gases given off by a mine fire. One-tenth of 1% may be fatal in ten minutes.

Working Face: Any place in a mine where material is extracted during a shift.

Reference: Kentucky Coal Council's Coal Education Web Site, Glossary of Mining Terms, www.coaleducation.org/glossary.htm (accessed 26 March 2004).

BIBLIOGRAPHY

Primary Sources
Abbreviations

Anglican Church of Canada, Calgary Diocese – ACCCD

Fernie & District Historical Society – FDHS

Glenbow Archives – GA

Provincial Archives of Alberta – PAA

Provincial Archives of British Columbia – PABC

Roman Catholic Diocese Archives – RCDA

Archival Sources

Aikins, C.B. (Oral History), 86.425 Provincial Archives of Alberta (PAA).

Anglican Church of Canada, Calgary Diocese fonds (ACCCD).

Annual Reports of the Minister of Mines for British Columbia, 1902–1928 (FDHS).

Attorney General (Alberta) – Fort Macleod District Court, Workman's Compensation Claims, 82.235 (PAA).

Attorney General (Alberta) – Inquest files, 67.172 and 68.261 (PAA).

Attorney General (British Columbia) – Inquisitions & Inquests, GR0429 and GR0431 (PABC).

Brown, H. Leslie, MS0736 (PABC).

Chuchla, Walter, M218, Glenbow Archives (GA).

Commission on Explosion in B North Mine, Coal Creek, 1915, GR0804 (PABC).

Crowsnest Resources, M1561 (GA).

Department of Labour, Economics & Research Branch – Strikes & Lockouts, GR1695 (PABC).

Drain, Charlie, M8645 (GA).

ERCB Mining Division – Mine Inspectors Reports, 77.237 (PAA).

    Hillcrest Collieries

    West Canadian Colliery Company

    International Coal and Coke Company

    Hillcrest Mohawk Colliery

    McGillivray Creek Colliery

Ethnic Research Project, M3553 (GA).

Fernie and District Historical Society (FDHS).

    Accident Record Book, Coal Creek Mine, September 1907–March 1914

    Bumps and Outbursts – Maps and Records

    Bumps and Outbursts – Correspondence, 1911–1944

    Funeral Records Book, 1920–1926

    Funeral Records, 1926–1938

    Transcripts of Oral History Interviews

Green, Raoul, M5764 (GA).

Gushul Family, M6163 (GA).

Hillcrest (miscellaneous), 75.198 (PAA).

Hutcheson, Sidney, MS0030 (PABC).

Hutchinson, William, M558 (GA).

Independent Order of Oddfellows, M5783 (GA).

Kennedy, Janet. *The McGillivray Creek Coal & Coke Co. and the International Coal & Coke Co., Coleman, Alberta: A Social History.* Unpublished research paper, 1984, M7742 (GA).

Legislative Assembly, 70.414 (PAA).

Public Works Annual Reports, later *Annual Reports of the Mines Branch of the Province of Alberta*, 68.149 (PAA).

Register regarding Investigation of Funeral Costs, 89.463 (PAA).

Report of Enquiry as to Funeral Costs in the Province of Alberta, February 11, 1932, Alex Ross to the Honorable Lieutenant Governor-General of Alberta, Sessional Paper No. 72

Roman Catholic Diocese Archives (RCDA).

    St. Cyril's – Bellevue

    Holy Spirit – Coleman

St. Theresa's – Hillcrest

St. Anne's – Blairmore

Speech (1916) by Edward Keith on 1914 Hillcrest Disaster, 86.67 (PAA).

Supreme Court – Probate/estate files, GR1458 and GR2519 (PABC).

Taped and Written Interviews, 72.355 (PAA).

Testimony given to Alberta Coal Commission 1907, 74.1 (PAA).

Tom Kirkham Oral History Project, RCT 260/RCT264 (GA).

United Mine Workers of America, M2239 (GA).

Vital Statistics, 87.385 (PAA).

    Hillcrest

    Blairmore

Newspapers

Bellevue Times

Blairmore Enterprise

Calgary Daily Herald

Coleman Bulletin

Coleman Journal

Coleman Miner

Fernie District Ledger

Fernie Free Press

Pincher Creek Echo

Toronto Globe

Legislation

An Act to make Regulations with respect to Coal to Mines, *Statutes of the Province of Alberta Passed on the First Session of the First Legislative Assembly*. Edmonton: King's Printer, 1906.

An Act Respecting Mines. *Revised Statutes of Alberta, 1922*. Edmonton: King's Printer, 1922.

An Act with respect to Compensation to Workmen for Injuries Suffered in the Course of Their Employment. *Statutes of the Province of Alberta Passed on the Third Session of the First Legislative Assembly*. Edmonton: King's Printer, 1908.

Secondary Sources

Articles and Chapters

———, "The Horse Driven Funeral in Early America." *The Driving Digest Magazine* 105 (1998): 8–12 and 22.

Adams, Sheila. "Women, Death and *In Memoriam* Notices in a Local British Newspaper." In *The Unknown Country: Death in Australia, Britain and the USA*, edited by Kathy Charmaz, Glennys Howarth, Allan Kellehear, 98–112. New York: St. Martin's Press, 1997.

Ashworth, James. "Firedamp: Hydrogen, Methane, Ethane and Propane, Plus Air in British Columbia." *Canadian Mining and Metallurgical Bulletin* 94 (February 1920): 125–46.

———. "Safety Lamps and Colliery Explosions." *Canadian Mining Review* 21, no. 6 (June 1902): 142–46.

———. "Safety Lamps and Colliery Explosions." *Canadian Mining Review* 21, no. 11 (November 1902): 267–69.

Badenhorst, J.C., and S.J. Van Schalkwyk. "Minimizing Post Traumatic Stress in Critical Mining Incidents." *Employee Assistance Quarterly* 7, no. 3 (1992): 79–90.

Beito, David T. "Thy Brother's Keeper: The Mutual Aid Tradition of American Fraternal orders." *Policy Review* 70 (Fall 1994): 55.

Benson, John. "Charity's Pitfalls: The Senghenydd Disaster." *History Today* 43 (November 1993): 7–9.

Blakemore, William. "Safety Lamps and Colliery Explosions" *Canadian Mining Review* 21, no. 9 (September 1902): 226–27.

—— "Some Recent Experiments in Blasting with Compressed Cartridges." *Journal of the Canadian Mining Institute* 1 (1898): 3–6.

Caragata, Warren. "The Labour Movement in Alberta: An Untold Story." In *Essays on the Political Economy of Alberta*, edited by David Leadbetter, 99–137. Toronto: New Hogtown Press, 1984.

Clawson, Mary Ann. "Fraternal Orders and Class Formation in the Nineteenth Century United States." *Comparative Studies in Society and History* 27, no. 4 (October 1985): 672–95.

Cowell, Daniel David. "Funerals, Family and Forefathers: a View of Italian-American Funeral Practices." *Omega* 16, no. 1 (1985–86): 69–85.

Crowell, Douglas L. "Death Underground: The Millfield Mining Tragedy." *Timeline* 14, no. 5 (September–October 1997): 42–56.

D'Alton, Phillip. "Prayers to Broken Stones: War and Death in Australia." In *The Unknown Country: Death in Australia, Britain and the USA*, edited by Kathy Charmaz, Glennys Howard, and Allan Kellehear, 45–57. New York: St. Martin's Press, 1997.

DeMont, John. "One Last Whistle." *Macleans* 114, no. 32 (6 August 2001): 16–19.

Dombroski, Wolf R. "Again and Again: Is a Disaster What We Call a Disaster?" In *What is a Disaster? Perspectives on the Question*, edited by E.L. Quarentelli, 19–30. London: Routledge, 1998.

Donovan, Arthur L. "Health and Safety in Underground Coal Mining, 1900–1969: Professional Conduct in a Peripheral Industry." In *The Health and Safety of Workers: Case Studies in the Politics of Professional Responsibility*, edited by Ronald Bayer, 72–138. New York: Oxford University Press, 1988.

Dvorak, Grace Arbuckle. "Childhood Remembered: A Coal Creek Memoir." In *The Forgotten Side of the Border: British Columbia's Elk Valley and Crowsnest Pass*, edited by Wayne Norton and Naomi Miller, 188–94. Kamloops: Plateau Press, 1998.

Edgette, J. Joseph. "The Epitaph and Personality Revelation." In *Cemeteries and Gravemarkers: Voices of American Culture*, edited by Richard E. Meyer, 87–101. Ann Arbor: UMI Research Press, 1989.

Emrich, Duncan. "Songs of the Western Miners." *California Folklore Quarterly* I (1942): 213–32.

Felske, Lorry W. "The Challenge Above Ground: Surface Facilities at Crowsnest Pass Mines before the First World War." In *A World Apart: The Crowsnest Communities of Alberta and British Columbia*, edited by Wayne Norton and Tom Langford, 158–69. Kamloops: Plateau Press, 2002.

Fritz, Paul S. "The Undertaking Trade in England: Its Origins and Early Development, 1660–1830." *Eighteenth-Century Studies* 28, no. 2 (Winter 1994–95): 241–53.

Gavin, James F., and Robert E. Kelley. "The Psychological Climate and Reported Well-Being of Underground Miners: An Exploratory Study." *Human Relations* 31, no. 7 (July 1978): 567–81.

Godfrey, E. "Explosives and Blasting in Coal Mining." *Bulletin of the Canadian Institute of Mining and Metallurgy* 114 (1921): 967–71.

Graham, Thomas. "Gaseous Mines in the Crow's Nest Pass Coal Field." *Coal Age* 10, no. 23 (1916): 920–23.

Gray, F.W. "Mine Rescue Work in Canada." *Canadian Mining Journal* (February 1913): 116–17.

Hershiser, Marvin R., and E.L. Quarantelli, "The Handling of the Dead in a Disaster." *Omega* 7, no. 3 (1976): 195–207.

Howarth, Glennys. "Professionalising the Funeral Industry in England 1700–1960." In *The Changing Face of Death: Historical Accounts of Death and Disposal*, edited by Peter C. Jupp, 120–34. London: MacMillan Press, 1997.

Kalisch, Philip A. "Death Down Below: Coal Mine Disasters in Three Illinois Counties, 1904–1962." *Journal of the Illinois State Historical Society* 65, no. 1 (Spring 1972): 5–21.

Keiser, John H. "The Union Miners Cemetery at Mt. Olive, Illinois: A Spirit-Thread of Labor History." *Journal of the Illinois State Historical Society* 62, no. 3 (Autumn 1969): 229–66.

Kinnear, John. "Fred Alderson: The Hosmer Hero." In *The Forgotten Side of the Border: British Columbia's Elk Valley and Crowsnest Pass*, edited by Wayne Norton and Naomi Miller, 177–84. Kamloops: Plateau Press, 1998.

Knowles, Norman. "'A Manly Commonsense Religion': Revivalism and the 1909 Kootenay Campaign in the Crowsnest Pass." In *A World Apart: The Crowsnest Communities of Alberta and British Columbia*, edited by Wayne Norton and Tom Langford, 3–13. Kamloops: Plateau Press, 2002.

Kreps, Gary A. "Disaster as Systemic Event and Social Catalyst." In *What is a Disaster? Perspectives on the Question*, edited by E.L. Quarentelli, 31–55. London: Routledge, 1998.

Lindemann, Erich. "Symptomatology and Management of Acute Grief." *American Journal of Psychiatry* 101 (1944): 141–48.

Lukasiewicz, Krystyna. "Polish Community in the Crowsnest Pass." *Alberta History* 36, no. 4 (Autumn 1988): 1–10.

Margolis, Eric. "Western Coal Mining as a Way of Life: An Oral History of the Colorado Coal Miners to 1914." *Journal of the West* 24, no. 3 (July 1985): 5–126.

McDonald, Duncan. "Mine Rescue and First Aid Work." *Bulletin of the Canadian Institute of Mining and Metallurgy* 115 (December 1921): 1048–52.

McEvoy, James. "Notes on Some Special Features of Coal Mining in the Crow's Nest." *Canadian Mining Review* 23, no. 3 (March 1904): 51–53.

Morgan, Wesley. "The One Big Union and the Crowsnest Pass." In *A World Apart: The Crowsnest Communities of Alberta and British Columbia*, edited by Wayne Norton and Tom Langford, 113–19. Kamloops: Plateau Press, 2002.

Oliver-Smith, Anthony. "What is a Disaster?" In *The Angry Earth: Disaster in Anthropological Perspective*, edited by Anthony-Oliver Smith and Susanna M. Hoffman, 1–33. New York: Routledge, 1999.

Ploeger, Andreas. "A 10-Year Follow up of Miners Trapped for 2 Weeks under Threatening Circumstances." *Stress and Anxiety* 4 (1977): 23–28.

Porfiriev, Boris N. "Issues on the Definition and Delineation of Disasters and Disaster Areas." In *What is a Disaster? Perspectives on the Question*, edited by E.L. Quarentelli, 56–72. London: Routledge, 1998.

Quarantelli, E.L., and Russell R. Dynes. "Community Conflict: Its Absence and Its Presence in Natural Disasters." *Mass Emergencies* 1 (1976): 139–52.

Quigley, J. Somerville. "Methods of Drawing Pillars in Pitching Seams." *Transactions of the Canadian Mining Institute (Canadian Institute of Mining Journal)* 17 (1914): 406–14.

Ratajczak, Zofia, and Marek Adamiec. "Some Dimensions in the Perception of Occupational Hazards by Coal Miners." *Polish Psychological Bulletin* 20, no. 3 (1989): 171–82.

Saad, Michael. "Mining Disasters and Rescue Operations at Michel Before World War Two." In *The Forgotten Side of the Border: British Columbia's Elk Valley and Crowsnest Pass*, edited by Wayne Norton and Naomi Miller, 127–32. Kamloops: Plateau Press, 1998.

Scandura, Jani. "Deadly professions: Dracula, Undertakers and the Embalmed Corpse." *Victorian Studies* 40, no. 1 (Autumn 1996): 1–30.

Seager, Alan. "Socialists and Workers: The Western Canadian Coal Miners, 1900–1921." *Labor/Le Travail* 16 (Fall 1985): 23–59.

Smith, Barbara. "I Know Them All: Monongah's Faithful Father Briggs." *Goldenseal* 25, no. 4 (Winter 1999): 21–24.

Smith, Frank B. "Coal Mining in the North-West Territories and its Probable Future." *Canadian Mining Review* 21, no. 4 (April 1902): 79–81.

———, and Raoul Green. "The Frank Disaster."*Canadian Mining Review* 22, no. 5 (May 1903): 102–3.

Stirling, John T. *The Bellevue explosions, Alberta, Canada: an account of, and subsequent investigation concerning, three explosions produced by sparks from falls of roof: a paper read before the Institution of Mining Engineers*. London: Institute of Mining Engineers, 1913.

———. "Coal Mining in Alberta during 1914." *Coal Age* 7, no. 2 (1915): 61–62.

Stout, Steve. "Tragedy in November: The Cherry Mine Disaster." *Journal of the Illinois State Historical Society* 72, no. 1 (February 1979): 57–69.

Strachan, Robert. "The Crowsnest Pass Coalfield." *Canadian Mining and Metallurgical Bulletin* 102 (October 1920): 767–77.

———. "Industrial Conditions in the Crowsnest Pass Coal-Field." *Canadian Mining and Metallurgical Bulletin* 148 (August 1924): 519–30.

Studenski, Ryszard, and Joanna Barcyzk. "Occupational Stressors in Mining as Related to Health, Job, Attitudes and Accident-Making." *Polish Psychological Bulletin* 18, no. 3 (1987): 159–68.

Ursano, Robert J., and James E. McCarrol. "Exposure to Traumatic Death: The Nature of the Stressor." In *Individual and Community Reponses to Trauma and Disaster: The Structure of Human Chaos*, edited by Robert J. Ursano, Brian G. McCaughey, and Carol S. Fullerton, 46–71. Cambridge: Cambridge University Press, 1994.

Watchempino, Lynette. "Dark as a Dungeon, Way Down in the Mine." *Suicide and Life-Threatening Behavior* 10, no. 4 (Winter 1980): 244–50.

Weeks, James L. "From Explosions to Black Lung: A History of Efforts to Control Coal Mine Dust." *Occupational Medicine* 8, no. 1 (January–March 1993): 1–17.

Weisaeth, Lars. "Psychological and Psychiatric Aspects of Technological Disasters." In *Individual and Community Responses to Trauma and Disaster: The Structure of Human Chaos*, edited by Robert J. Ursano, Brian G. McCaughey, and Carol S. Fullerton, 72–102. Cambridge: Cambridge University Press, 1994.

Yarmie Andrew. "Community and Conflict: The Impact of the 1902 Explosion at Coal Creek." In *The Forgotten Side of the Border: British Columbia's Elk Valley and Crowsnest Pass*, edited by Wayne Norton and Naomi Miller, 195–205. Kamloops: Plateau Press, 1998.

Books

Adams, James Taylor. *Death in the Dark: A Collection of Factual Ballads of American Mine Disasters*. N.P.: Adams-Mullins Press, 1941.

Adams, R. *Eternal Prairie: Exploring Rural Cemeteries of the West*. Calgary: Fifth House Publishers, 1999.

Anderson, Frank. *Tragedies of the Crowsnest Pass*. Surrey: Frontier Books, 1983.

Ashworth, James. *Safety lamps and colliery explosions*. Canadian Mining Institute: Published by the Authority of Council at the Office of the Secretary, Ottawa, 1902.

Avery, Donald. *"Dangerous Foreigners:" European Immigrant Workers and Labour Radicalism in Canada, 1896–1932*. Toronto: McClelland and Stewart, 1979.

Benet, Sula. *Song, Dance and Customs of Peasant Poland*. London: Dennis Dobson, 1951.

Bennett, Glin. *Beyond Endurance: Survival at the Extremes*. New York: St. Martin's Press, 1983.

Bercuson, David J. *Fools and Wise Men: The Rise and Fall of the One Big Union*. Toronto: McGraw-Hill Ryerson, 1978.

Braithwate, John. *To Punish or Persuade: Enforcement of Coal Mine Safety*. Albany: State University of New York Press, 1985.

Brown, Roger David. *Blood on the Coal: The Story of the Springhill Mining Disasters.* Windsor: Lancelot Press, 1976.

Byrne, M.B. Venini. *From the Buffalo to the Cross: A History of the Roman Catholic Diocese of Calgary.* Calgary: Calgary Archives and Historical Publishers, 1973.

Callaway, C.F. *The Art of Funeral Directing: A Practical Manual on Modern Funeral Directing Methods.* Chicago: Undertakers' Supply Co., 1928.

Cerney, Hazel. *The Silent Hills: the Kropinak family of Chapel Rock.* Blairmore: H.K. Cerney, 1992.

Chrismas, Lawrence. *Alberta Miners: A Tribute.* Calgary: Cambria Publishing, 1993.

———. *Coal Dust grins: portraits of Canadian Coal Miners.* Calgary: Cambria Publishing, 1998.

The Coal Mining Industry in the Crow's Nest Pass. Compiled by Sharon Babaian from research papers by Lorry Felske. Edmonton: Alberta Culture, 1985.

Coleman's Fiftieth Anniversary Booklet. Lethbridge: Lethbridge Herald, 1953.

Comish, Shaun. *The Westray Tragedy: A Miners Story.* Halifax: Fernwood Publishing, 1993.

Crow's Nest Pass Coal Company. *Special Rules to be Observed by the Owners, Managers, Overmen, Master Mechanics, Firebosses & Workmen of the Collieries of the Crow's Nest Pass Coal Company, Limited.* c1897.

Crowsnest Pass Historical Society. *Crowsnest and its People,* Coleman: Crowsnest Pass Historical Society, 1979.

The Cyclopaedia of Fraternies. Edited by Albert C. Stevens. New York: E.B. Treat and Company, 1907.

Davies, Douglas J. *Death, Ritual and Belief: The Rhetoric of Funerary Rites.* London: Cassell Publishers, 1997.

Dick, W.J. *Mine-Rescue Work in Canada.* Ottawa: Rolls L. Crain Co., 1912.

Duke, David C. *Writings and Miners: Activism and Imagery in America.* Lexington: University of Kentucky Press, 2002.

Erikson, Kai T. *Everything in Its Path: Destruction of Community in the Buffalo Creek Flood.* New York: Simon and Schuster, 1976.

Fernie Free Press. *The Crow's Nest Pass Illustrated Souvenir Edition of the Fernie Free Press.* Fernie: Fernie Free Press, 1907.

Fernie Historical Association. *Backtracking with the Fernie Historical Association.* Fernie: Fernie Historical Association, 1967.

Freemasons Grand Lodge of Alberta. *Constitution of the Grand Lodge of Alberta, Ancient, Free and Accepted Masons as Adopted at the Special Communication, held at Calgary, February 25th and 26th, 1919.*

Fritz, Charles E., and J.H. Mathewson. *Convergence Behavior in Disasters: A Problem in Social Control.* Disaster Study Number 9. Washington: National Research Council, 1957.

Gilpin, Edwin. *The Use of Safe Explosives in Mines: Part II, The Results of the Experiments.* Transactions of the Canadian Society of Civil Engineers, 1892.

Goodrich, Carter. *The Miner's Freedom: A Study of the Working Life in a Changing Industry.* Boston: Marshall Jones Co., 1925.

Grand Lodge of Ontario Freemasons. *Ceremonies of Organizing, Constituting and Consecrating a New Lodge.* 1893.

Green, Archie. *Only a Miner: Studies in Recorded Coal-Mining Songs.* Chicago: University of Illinois Press, 1972.

Greene, Melissa Faye. *Last Man Out: The Story of the Springhill Mine Disaster.* Toronto: Harcourt, 2003.

Gunderson, Harald. *A History of Funeral Service in Alberta: Sixty Five Years of Serving Albertans, 1928–1993.* Calgary: Alberta Funeral Service Association, 1993.

Habenstein, Robert W., and Willima M. Lamers. *The History of American Funeral Directing.* Revised ed. Milwaukee: Bulfin Printers, 1962.

Harger, John. *Coal and the Prevention of Explosions and Fires in Mines*. Newcastle-Upon-Tyne: Andrew Reid and Co., 1913.

Harrell, W. Andrew. *Accident History, Perceived Risk of Personal Injury, and Job Mobility as Factors Influencing Occupational Accident Fatalism*. Edmonton Area Series Report No. 46. Edmonton: University of Alberta Centre for Experimental Sociology, 1985.

Harrington, M.A. *Twenty-Fifth Anniversary, St. Anne's Parish, Blairmore, Alberta*. Blairmore: Enterprise Job Print, 1935.

Harvey, Horace. *Report of the Inquiry into an Explosion which occurred on the 23rd Day of November, 1926 in a Coal Mine Operated by the McGillivray Creek Coal and Coke Company at Coleman Alberta*. 10 December 1927.

Hodgkinson, Peter E., and Michael Stewart. *Coping with Catastrophe: A Handbook of Post-Disaster Psychosocial Aftercare*. 2nd ed. London: Routledge, 1998.

Hutcheson, Sydney. *Depression Stories*. Vancouver: New Star Books, 1976.

Independent Order of Oddfellows, Esther Rebekah Lodge, No. 20 (Fernie, B.C.). *Constitution, By-Laws and Rules of Order of Esther Rebekah Lodge No. 20, IOOF instituted at Fernie, BC, April 5, 1907*

Independent Order of Oddfellows, Mount Fernie Lodge No. 47. *Constitution, By-Laws and Rules of Order of Mount Fernie Lodge No. 47, IOOF under Jurisdiction of the Grand Lodge of British Columbia, Fernie, BC, Instituted February 24th, 1902*.

Iserson, Kenneth V. *Grave Words: Notifying Survivors about Sudden, Unexpected Deaths*. Tucson: Galen Press, 1999.

Kearl, Michael C. *Endings: A Sociology of Death and Dying*. Oxford: Oxford University Press, 1989.

Kerr, J. William. *Frank Slide*. Priddis: Barker Publishing, 1990.

Korson, George. *Coal Dust on the Fiddle: Songs and Stories of the Bituminous Industry*. 2nd ed. Hatboro: Folklore Associates, 1965.

Levy, Elizabeth, and Tad Richards. *Struggle and Lose, Struggle and Win: The United Mine Workers*. New York: Four Winds Press, 1977.

Lloyd, A.L. *Come All Ye Bold Miners: Ballads and Songs of the Coalfields*. 2nd ed. London: Lawrence and Wishart, 1978.

Long, Priscilla. *Where the Sun Never Shines: A History of America's Bloody Coal Industry*. New York: Paragon House, 1989.

Lucas, Rex A. *Men in Crisis: A Study of a Mine Disaster*. New York: Basic Books, 1969.

Lynch, Thomas. *The Undertaking: Life Studies from the Dismal Trade*. New York: W.W. Norton and Co., 1997.

Memoirs of Polish Immigrants in Canada. Edited by Benedykt Heydenkorn. Toronto: Canadian-Polish Research Institute, 1979.

Millar, Nancy. *Once Upon a Tomb: Stories from Canadian Graveyards*. Calgary: Fifth House Publishers, 1997.

———. *Remember Me as You Pass By: Stories from Prairie Graveyards*. Calgary: Glenbow Alberta, 1994.

Miller, Joan. *Aberfan: A Disaster and its Aftermath*. London: Constable and Co., 1974.

Morrow, R.A.H. *Story of the Springhill Disaster*. St. John: R.A.H. Morrow, 1891.

Muise, Delphin A., and Robert G. McIntosh. *Coal Mining in Canada: A Historical and Comparative Overview*. Ottawa: National Museum of Science and Technology, 1996.

O'Donnell, John C. *"And Now the Fields are Green": A Collection of Coal Mining Songs in Canada*. Sydney: University College of Cape Breton Press, 1992.

Norris, John. *Strangers Entertained: A History of the Ethnic Groups of British Columbia*. Vancouver: Evergreen Press, 1971.

Palmer, Howard. *Land of the Second Chance: A History of Ethnic Groups in Southern Alberta*. Lethbridge: Lethbridge Herald, 1972.

Perrow, Charles. *Normal Accidents: Living with High-Risk Technologies*. Princeton: Princeton University Press, 1999.

Polish Settlers in Alberta: Reminiscences and Biographies. Edited by Joanna Matejko. Toronto: Polish Alliance Press, 1979.

Radecki, Henry. *Ethnic Organizational Dynamics: The Polish Group in Canada.* Waterloo: Wilfrid Laurier University Press, 1979.

Rejman, Jerome. *Holy Spirit Catholic Church 1905–1980: A History of the Holy Spirit Parish Coleman, Alberta* (1980).

Richard, Justice K. Peter. *The Westray Story: A Predictable Path to Disaster.* Report of the Westray Mine Public Inquiry. 3 vols. Nova Scotia, November 1997.

Robertson, W.F. *Reports on the Fernie Coal Mine Explosion.* Victoria: King's Printer, 1902.

Ross, Toni. *Oh! The Coal Branch.* Edmonton: Mrs. Toni Ross, 1974.

Runnalls, Reverend F.E. *It's God's Country: A Review of the United Church and its Founding Partners, the Congregational, Methodist and Presbyterian Churches in British Columbia.* N.P., 1974.

Ryan, Judith Hoegg. *Coal in Our Blood: 200 Years of Coal Mining in Nova Scotia's Pictou County.* Halifax: Formac Publishing, 1992.

Seltzer, Curtis. *Fire in the Hole: Miners and Managers in the American Coal Industry.* N.P.: University Press of Kentucky, 1985.

Silverman, Eliane Leslau. *The Last Best West: Women on the Alberta Frontier, 1880–1930.* 2nd ed. Calgary: Fifth House Publishers, 1998.

Sloane, David Charles. *The Last Great Necessity: Cemeteries in American History.* Baltimore: Johns Hopkins University Press, 1991.

Takamiya, Makoto. *Union Organization and Militancy: Conclusions from a Study of the United Mine Workers of America, 1940–1974.* Berlin: Verlag Anton Hain – Meisenheim am Glan, 1978.

Taylor, A.J.W. *Disasters and Disaster Stress.* New York: AMS Press, 1989.

Wentworth, Bert, and Harris Hawthorne Wilder. *Personal Identification: Methods for the Identification of Individuals, Living or Dead.* Chicago: T.G. Cooke, 1932.

Wood, Patricia K. *Nationalism from the Margins: Italians in Alberta and British Columbia.* Montreal: McGill-Queen's University Press, 2002.

Wright, Bob. *Sudden Death: A Research Base for Practice.* New York: Churchill Livingstone, 1996.

Unpublished Sources

Felske, Lorry William. "Studies in the Crow's Nest Pass Coal Industry from its Origins to the End of World War I." Ph.D. diss., University of Toronto, 1991.

Green, Julie Anne. "Calculated Risks: Worker, Owner and Government Attitudes Toward Safety in the Crow's Nest Pass Coal Mines, 1900–1915." Master's thesis, University of Calgary, 1990.

Karas, Frank Paul. "Labour and Coal in the Crowsnest Pass: 1925–1935." Master's thesis, University of Calgary, 1972.

Lenfesty, Corrine B. "Choices for the Living, Honour for the Dead: A Century of Funeral and Memorial Practices in Lethbridge." Master's thesis, University of Lethbridge, 1998.

Michrina, Barry Paul. "Lives of Dignity, Acts of Emotion: Memories of Central Pennsylvania Bituminous Coal Miner Families." Ph.D. diss., State University of New York at Binghamton, 1992.

Monteressi, Christopher. "Emotional Hijacking Versus Safe Behavior." Ph.D. diss., College of Engineering and Mineral Resources at West Virginia University, 1998.

Munn Gafuik, Jo-Ann. "Bishops, Priests and Immigrants: The Roman Catholic Diocese of Calgary and the Immigrant Question." Master's thesis, University of Calgary, 1996.

Other

Blood on the Coal [video recording]. Toronto: CBC International Sales, 1999.

Crowsnest Pass Historical Society, unpublished museum notes on Hillcrest Explosion, copied August 2001.

Hillcrest Mine Site Educational Program: "Hear That Whistle Blow". Frank Slide Visitor Centre, 2001.

'Hillcrest Mine,' by James Keelaghan, in *Small Rebellions* [sound recording], 1990.

Interview with Roy Lazarrotto, Bellevue Mine, 11 November 2001.

Lethbridge Funerals Exhibit, Galt Museum and Archives, June 2001

Cemeteries: Bellevue, Blairmore Union, Blairmore Roman Catholic, Coleman Union, Coleman Roman Catholic, Elk Valley/Michel, Hillcrest, Hosmer, Fernie, Pincher Creek.

Electronic Sources

'Coal: An Introduction' (describing the 'coalification' process. University of Newcastle, http://www.newcastle.edu.au/department/gl/cfkd/undp.htm (accessed 5 December 2001).

Crowsnest Coal: Mining Disasters, http://www.crowsnest.bc.ca/coal09.html (accessed 26 March 2004).

Crowsnest Pass Railway Route, http://www.crowsnest.bc.ca/fernie01.html (accessed 26 March 2004.)

Glossary of Mining Terms, Kentucky Coal Council Coal Education, http://www.coaleducation.org/glossary.htm (accessed 26 March 2004).

'Memorial to Victims of the Cocoanut Grove Fire,' Find a Grave, http://www.findagrave.com/cgi-bin/fg.cgi?page=gr@GRid=1215 (accessed 26 March 2004).

Toledo Area Disaster Medical Assistance Team (TADMAT), "Blast Injuries" by Nick Olovos, *http://mediccom.org/public/tadmat/training/ndms/Blast_Document.pdf* (accessed 26 March 2004).

NOTES

Notes to Introduction

1. *Fernie Free Press*, 5 April 1907.

2. Mr. Heathcote, Chief Inspector of Mines, to Hon. W.H. Cushing, M.P., 30 April 1907, Mine Inspector's Reports, box 16, file 88a, Provincial Archives of Alberta 77.237 (hereinafter Reports PAA).

3. Ibid.

4. Historically, the Crowsnest Pass area has been variously called "Crow's Nest Pass" or "Crows Nest Pass." For the purposes of this publication, the modern denotation of Crowsnest Pass is used.

5. Crowsnest Pass Historical Society, *Crowsnest and its People* (Coleman: Crowsnest Pass Historical Society, 1979), 303; Coleman Local 2633 Minutes 1903–1909, 25 June 1906, United Mine Workers of America, Glenbow Archives M2239 (hereafter United Mine Workers GA).

6. Grace Arbuckle Dvorak, "Childhood Remembered: A Coal Creek Memoir," in *The Forgotten Side of the Border: British Columbia's Elk Valley and Crowsnest Pass*, ed. Wayne Norton and Naomi Miller (Kamloops: Plateau Press, 1998), 193.

7. Lawrence Chrismas, *Coal Dust Grins: Portraits of Canadian Coal Miners* (Calgary: Cambria Publishing, 1998), 74.

8. Interview with Harry Gate, Coleman 1927, PAA 72.355, Taped and written interviews, Item 4.

9. *Crowsnest and Its People*, 150.

10. Allen Seager, "Socialists and Workers: The Western Canadian Coal Miners, 1900–1921," *Labour/Le Travail* 16 (Fall 1985): 33; Bruce Ramsay, *100 Years of Coal Mining: The Elk River Valley* (Sparwood: Ramsay Publications, 1997), 160.

11. *Crowsnest and Its People*, 151 and 756.

12. W.F. Robertson, *Reports on the Fernie Coal Mine Explosion* (Victoria: King's Printer, 1902).

13. *Fernie Free Press*, 26 May 1902.

14. *Crowsnest and Its People*, 330.

15. *Coleman Miner*, 26 June 1908.

16. Frank Slide Interpretative Centre, *Cultural Diversity* display panel, 3 May 2003.

17. Bumps and Outbursts – Maps and Records, R. Strachan, Fernie and District Historical Society (hereinafter Bumps Maps FDHS).

18. Duncan McDonald, "Mine Rescue and First Aid Work," *Bulletin of the Canadian Institute of Mining and Metallurgy* 115 (December 1921): 1048. W.J. Dick, *Mine-Rescue Work in Canada* (Ottawa: Rolls L. Crain Co., 1912), 24–26.

19. Oral History of C.B. Aikens, Recorded 25 August 1986, PAA 86.425.

20. *Blairmore Enterprise*, 2 December 1926.

21. Eric Margolis, "Western Coal Mining as a Way of Life: An Oral History of the Colorado Coal Miners to 1914", *Journal of the West* 24, no. 3 (July 1985): 6.

22. Ibid.

23. Rex Lucas, *Men in Crisis: A Study of a Mine Disaster* (New York: Basic Books, 1969), 5–6.

24. See especially George Korson, *Coal Dust on the Fiddle: Songs and Stories of the Bituminous Industry*, 2nd ed. (Hatoboro: Folklore Associates Inc, 1965); John C.O'Donnell, *"And Now the Fields are Green": A Collection of Coal Mining Songs in Canada* (Sydney: University College of Cape Breton Press, 1992); A.L. Lloyd, *Come All Ye Bold Miners: Ballads and Songs of the Coalfields* 2nd ed. (London: Lawrence and Wishart, 1978). All these collections, whether collected in the United States or Canada, show remarkable homogeneity in the types of songs chosen and the various editions of the same songs handed down through the generations.

25. O'Donnell, "*And Now the Fields are Green*," ix.

26. *Crowsnest and Its People*, 390.

27 Ballads of the Avondale, Lick Branch, and Fraterville mine disasters are found in James Taylor Adams, *Death in the Dark: A Collection of Factual Ballads of American Mine Disasters* (n.p.: Adams-Mullins Press, 1941), 19, 29, 93. The Cherry mine disaster narrative is found in George Korson, *Coal Dust on the Fiddle*, 252.

28 Archie Green, *Only a Miner: Studies in Recorded Coal-Mining Songs* (Chicago: University of Illinois Press, 1972), 113.

29 Erich Lindemann, "Symptomatology and Management of Acute Grief," *American Journal of Psychiatry* 101 (1944): 141–48.

30 See in particular Lucas, *Men in Crisis*; Andreas Ploeger, "A 10-Year Follow up of Miners Trapped for 2 Weeks under Threatening Circumstances," *Stress and Anxiety* 4 (1977): 23–28; J.C. Badenhorst and S.J. Van Schalkwyk, "Minimizing Post Traumatic Stress in Critical Mining Incidents," *Employee Assistance Quarterly* 7, no. 3 (1992): 79–90; Ryszard Studenski and Joanna Barcyzk, "Occupational Stressors in Mining as Related to Health, Job, Attitudes and Accident-Making," *Polish Psychological Bulletin* 18, no. 3 (1987): 159–68; Zofia Ratajczak and Marek Adamiec, "Some Dimensions in the Perception of Occupational Hazards by Coal Miners," *Polish Psychological Bulletin* 20, no. 3 (1989): 171–82.

31 See in particular Joan Miller, *Aberfan: A Disaster and its Aftermath* (London: Constable and Co., 1974); and Kai T. Erikson, *Everything in its Path: Destruction of a Community in the Buffalo Creek Flood* (New York: Simon and Schuster, 1976).

32 The Committee on Disaster Studies was created in 1952 in the United States, followed by the Disaster Research Group, the National Opinion Research Centre and the Office of Emergency Preparedness. A.J.W. Taylor, *Disasters and Disaster Stress* (New York: AMS Press, 1989), 57.

33 Erikson, *Everything in its Path*, 165.

34 Ibid., 153.

35 Phillip D'Alton, "Prayers to Broken Stones: War and Death in Australia," in *The Unknown Country: Death in Australia, Britain and the USA*, eds. Kathy Charmaz, Glennys Howard and Allan Kellehear (New York: St. Martin's Press, 1997), 49. While the reference here is to the Australian "digger" of ANZAC and Gallipoli fame, a parallel can be drawn between the bravery and courage attributed to both the diggers and the miners of the deeps. The media has reinforced images of both.

36 *Calgary Daily Herald*, 20 June 1914.

37 Arthur L. Donovan, "Health and Safety in Underground Coal Mining, 1900–1969: Professional Conduct in a Peripheral Industry," in *The Health and Safety of Workers: Case Studies in the Politics of Professional Responsibility*, ed. Ronald Bayer (New York: Oxford University Press, 1988), 94.

38 *District Ledger*, 7 February 1906.

39 David C. Duke, *Writers and Miners: Activism and Imagery in America* (Lexington: University of Kentucky Press, 2002), 3.

40 *Fernie Free Press*, 25 March 1904.

41 *Crowsnest and Its People*, 684.

42 John DeMont, "One Last Whistle," *Macleans* 114, no. 32 (August 2001): 17.

43 Lawrence Chrismas, *Alberta Miners: A Tribute* (Calgary: Cambria Publishing, 1993); and Chrismas, *Coal Dust Grins*.

44 Barry Paul Michrina, "Lives of Dignity, Acts of Emotion: Memories of Central Pennsylvania Bituminous Coal Miner Families," Ph.D. diss., State University of New York at Binghamton, 1992, 114–15.

Notes to Chapter 1

1 Quoted in A.L. Lloyd, *Come All Ye Bold Miners: Ballads and Songs of the Coalfields* (London: Lawrence and Wishart, 1978), 151.

2 James L. Weeks, "From Explosions to Black Lung: A History of Efforts to

Control Coal Mine Dust," *Occupational Medicine* 8, no. 1 (January–March 1993): 11.

3   W.J. Dick, *Mine-Rescue Work in Canada* (Ottawa: Rolls L. Crain Co., 1912), 7; see also John Braithwate, *To Punish or Persuade: Enforcement of Coal Mine Safety* (Albany: State University of New York Press, 1978). Brathwaite states on p. 1 that the explosion killed 1,100, while another reference states that the Courrieres mine explosion killed 1,230 men. See Arthur L. Donovan, "Health and Safety in Underground Coal Mining, 1900–1969: Professional Conduct in a Peripheral Industry," in *The Health and Safety of Workers: Case Studies in the Politics of Professional Responsibility*, ed. Ronald Bayer (New York: Oxford University Press, 1988), 110.

4   For discussions on the establishment of the companies on the British Columbia side of the border, see Robert Strachan, "The Crowsnest Pass Coalfield," *Canadian Mining and Metallurgical Bulletin* 102 (October 1920): 767; *The Coal Mining Industry in the Crow's Nest Pass*, compiled by Sharon Babaian from research papers by Lorry Felske (Edmonton: Alberta Culture, 1985), 10 (hereinafter "Babaian"); Robert Strachan, "Industrial Conditions in the Crowsnest Pass Coal-Field," *Canadian Mining and Metallurgical Bulletin* 148 (August 1924): 519; Thomas Graham, "Gaseous Mines in the Crow's Nest Pass Coal Field," *Coal Age* 10, no. 23 (December 1916): 920; Fernie Free Press, *The Crow's Nest Pass Illustrated Souvenir Edition of the Fernie Free Press* (Fernie: Fernie Free Press, 1907); Fernie Historical Association, *Backtracking with the Fernie Historical Association* (Fernie: Fernie Historical Association, 1967), 41, 53, 59.

5   For discussions on the establishment of the companies on the Alberta side of the border, see Crowsnest Pass Historical Society, *Crowsnest and its People* (Coleman: Crowsnest Pass Historical Society, 1979), 47, 129, 213; *Coleman's Fiftieth Anniversary Booklet* (Lethbridge: Lethbridge Herald, 1953), 25–26; Babaian, *The Coal Mining Industry in the Crow's Nest Pass*, 13.

6   The data for these charts is taken from the *Annual Reports* to the Minister of Mines for both Alberta and British Columbia. Alberta did not start producing statistics until 1906.

7   *Bellevue Times*, 30 December 1910.

8   Dick, *Mine-Rescue Work in Canada*, v.

9   Lorry Felske "Studies in the Crow's Nest Pass Coal Industry from its Origins to the End of World War 1," Ph.D. diss., University of Toronto, 1991, 198.

10 James McEvoy, "Notes on Some Special Features of Coal Mining in the Crow's Nest," *Canadian Mining Review* 23, no. 3 (March 1904): 51.

11 For a comprehensive look at the dangers of mining in the Crowsnest Pass, see Felske's "Studies in the Crow's Nest Pass," Chapter Four – "Crowsnest Pass Geology – Living with the Lethal Environment." Contemporary books and articles also dealt with various aspects of coal mining dangers, including John Harger, *Coal, and the Prevention of Explosions and Fires in Mines* (Newcastle-Upon-Tyne: Andrew Reid and Co., 1913); Graham, "Gaseous Mines in the Crow's Nest Pass Coal Field;"; William Blakemore, "Some Recent Experiments in Blasting with Compressed Cartridges," *Journal of the Canadian Mining Institute* 1 (1898): 3–6.

12 McEvoy, 52. Coal Creek dipped from 15 to 20 degrees; Morrissey from 8 to 24 degrees; Michel from 15 to 35 degrees.

13 General Manager of West Canadian Collieries Ltd. to John Stirling, Provincial Inspector of Mines, 13 May 1910, Mine Inspector's Reports, file 64a, PAA 77.237 (hereinafter Reports PAA) .

14 Babaian, *Coal Mining Industry*, 40.

15 Edwin Gilpin, *The Use of Safe Explosives in Mines: Part II, The Results of the Experiments* (Canadian Society of Civil Engineers, 1892).

16 *Report of the Special Commission Appointed to Inquire into the Causes of Explosion in Coal Mines*, 1903, Crowsnest Resources, box 14, file 209,

NOTES   167

Glenbow Archives M1561 (hereinafter Crowsnest Resources GA), 3–4.

17 E. Godfrey, "Explosives and Blasting in Coal Mining," *Bulletin of the Canadian Institute of Mining and Metallurgy* 114 (November 1921): 968–69. For a description of how a shot was fired, see Horace Harvey, *Report of the Inquiry into an Explosion which occurred on the 23rd Day of November, 1926 in a Coal Mine Operated by the McGillivray Creek Coal and Coke Company at Coleman Alberta*, 10 December 1927, 5–7.

18 Julie Anne Green, "Calculated Risks: Worker, Owner and Government Attitudes Toward Safety in the Crow's Nest Pass Coal Mines, 1900–1915," Master's thesis, University of Calgary, 1990, 19, 97–98.

19 "Coal: An Introduction" (describing the "coalification" process, University of Newcastle), *http://www.newcastle.edu.au/department/gl/cfkd/undp.htm* (accessed 5 December 2001).

20 Curtis Seltzer, *Fire in the Hole: Miners and Managers in the American Coal Industry*, (n.p.: University Press of Kentucky, 1985), 1–2.

21 Graham, "Gaseous Mines in the Crow's Nest Pass Coal Field;" 922. An analysis of the gas at the Michel Colliery No. 3 East Mine at Morrissey in 1916 gave volumes varying from a low of 1032 cubic feet to a high of 15,695 cubic feet for an average of 4,135 cubic feet per ton of coal mined. At the time of analysis, the No. 3 mine was producing 600 tons of coal per day and 3,500 cubic feet of gas. This daily volume is sufficient to heat fifteen to twenty average homes in Calgary for an entire year.

22 Strachan, "The Crowsnest Pass Coalfield," 767–68.

23 Robert Strachan, Senior Inspector of Mines to D. Harrington, Supervising Mining Engineer, Bureau of Mines, 8 December 1923, Bumps and Outbursts, Correspondence 1911–1944, Fernie and District Historical Society (hereinafter Bumps Correspondence FDHS).

24 Justice K. Peter Richard, *The Westray Story: A Predictable Path to Disaster: Report of the Westray Mine Public Inquiry*, 3 vols. (Nova Scotia, 1997) 1:202. Firedamp could be used to describe the air mixture, the ignition of methane or the methane itself. Note 59 on p. 202 reads: "Firedamp is a common British term for a flammable methane-air mixture, descriptive of the gas's potential to ignite or explode. It is also used as a synonym for methane."

25 W.F. Robertson, *Report on the Fernie Coal Mine Explosion* (Victoria: King's Printer, 1902).

26 Richard, *Westray Story* 1:279–80.

27 Leslie H. Brown, *The Pass: Biography of a District*, Provincial Archives of British Columbia MS0736 (hereinafter noted as Brown ms. PABC), 211.

28 For a thorough examination of the resistance to changing to safety lamps, see Julie Anne Green, "Calculated Risks," in particular 77, 87–94. Contemporary correspondence on the change to safety lamps is voluminous. For Alberta government reports on the problem, see the *Annual Reports of the Mines Branch of the Province of Alberta* (hereinafter *Alberta Mines Annual Report*). For correspondence between mine managers and government officials on the change, see Reports PAA. For example, see Memo for Mr. Stocks from Norman Fraser, 7 October 1908, file 40j, detailing a three-year list of complaints, mostly regarding safety lamps, between C.P. Hill of the Hillcrest mine and Mr. Heathcote, District Inspector of Mines. In British Columbia, the change from the Wolf Lamp to the new electric Edison lamp alleviated some of these problems. For example, *Annual Report to the Minister of Mines for British Columbia* (hereinafter *British Columbia Mines Annual Report*), *1917*, 426. Comments in the 1918 *Annual Report* showed that the men were more favourable to the change owing to amount of light shed by the Edison lamp. For contemporary articles on the issue, see William Blakemore, "Safety Lamps and Colliery Explosions," *Canadian Mining Review* 21, no. 9 (September 1902); James Ashworth, "Safety Lamps and Colliery Explosions," *Canadian Mining Review* 21, no. 6 (June 1902); James Ashworth, "Safety Lamps

and Colliery Explosions," *Canadian Mining Review* 21, no. 11 (November 1902); James Ashworth, *Safety Lamps and Colliery Explosions* (Ottawa: Canadian Mining Institute, 1902).

29 Elijah Heathcote, District Inspector of Mines to John Stocks, Deputy Minister of Public Works, 31 October 1908, file 40j, Reports PAA.

30 John T. Stirling and John Cadman, "The Bellevue explosions, Alberta, Canada: an account of, and subsequent investigation concerning, three explosions produced by sparks from falls of roof" (paper read before the Institution of Mining Engineers, London, 1913).

31 *British Columbia Mines Annual Report 1918*, 347. See also Robert Strachan, Senior Inspector of Mines to the Hon. William Sloan, Minister of Mines for British Columbia, 28 November 1922, Bumps Correspondence FDHS.

32 *Alberta Mines Annual Report 1909*, 113.

33 Strachan to Harrington, 8 December 1923, Bumps Correspondence FDHS.

34 Glossary of Mining Terms, Kentucky Coal Council Coal Education, *http://www.coaleducation.org/glossary.htm* (accessed 26 March 2004).

35 Jack McAndrew commentary, *Blood on the Coal* [video recording] (Toronto: CBC International Sales, 1999) (hereinafter *Blood on the Coal video*).

36 F.H. Shepherd, Blowouts of Gas, Morrissey British Columbia 1909, Bumps Correspondence FDHS.

37 Charles Perrow, *Normal Accidents: Living with High-Risk Technologies* (Princeton: Princeton University Press, 1999), 245.

38 Ibid.

39 William J. Kerr, *Frank Slide* (Priddis: Barker Publishing, 1990), 8.

40 *Crowsnest and Its People*, 557.

41 "Alphabetical List of Miners Imported from Great Britain," box 7, file 112, Crowsnest Resources GA.

42 Series IX Employee Records, files 215–238, Crowsnest Resources GA.

43 *Crowsnest and Its People*, 560, 586.

44 Lawrence Chrismas, *Alberta Miners: A Tribute* (Calgary: Cambria Publishing, 1993), 79.

45 *Crowsnest and Its People*, 554.

46 Report from District Inspector of Mines Moses Johnson to John T. Stirling, Chief Inspector of Mines, box 14, file 87e, Reports PAA.

47 The Phillips are buried in the Elk Valley Union Cemetery; the Puckeys and Munkwitzes are buried in the Fernie Cemetery. See also 12 December 1924 and 17 June 1910 *Fernie Free Press* and accounts of the 1917 Coal Creek explosion in the *Fernie Free Press* and *District Ledger*.

48 Annual Report for the Year Ending the 31st of December 1898, William Blakemore General Manager to the Directors of the Crows Nest Pass Coal Co. Ltd., 18 February 1899, file 20, Crowsnest Resources GA.

49 Series IX Employee Records, files 215–509, Crowsnest Resources GA.

50 Article on "The Hillcrest Mine Disaster" as told to Stanley W. Frolick by his father, a survivor of the 1914 mine explosion, PAA 75.198, Hillcrest (miscellaneous) (hereinafter Hillcrest misc. PAA), 33.

51 David J. Bercuson, *Fools and Wise Men: The Rise and Fall of the One Big Union* (Toronto: McGraw-Hill Ryerson, 1978), 11. The drawback to this method was that the more experienced could lose wages from having to spend time teaching the less experienced miner.

52 Carter Goodrich, *The Miner's Freedom: A Study of the Working Life in a Changing Industry* (Boston: Marshall Jones Company, 1925), 56, 60.

53 J. Somerville Quigley, "Methods of Drawing Pillars in Pitching Seams," *Transactions of the Canadian Mining Institute* 17 (1914): 406.

54 Brown ms. PABC, 200.

55 Article on Charlie Drain by Greg McIntyre, Herald Staff Writer, n.d., Charlie Drain, GA M8645 (hereinafter Drain GA).

56. Albert Goodwin Interview Tape, Tom Kirkham Oral History Project, GA RCT264-1 (hereinafter Oral History GA).

57. Lawrence Chrismas, *Coaldust Grins: Portraits of Canadian Coal Miners* (Calgary: Cambria Publishing, 1998), 73.

58. Gino Busato Interview, Transcripts of Fernie Interviews, FDHS (hereinafter Fernie Interviews FDHS).

59. Alrik Tiberg, Oral History GA.

60. *Crowsnest and Its People*, 524.

61. Article on "The Hillcrest Mine Disaster," Hillcrest misc. GA, 31.

62. Pasquale and Maria Perri, Fernie Interviews FDHS.

63. Toni Ross, *Oh! The Coal Branch* (Edmonton: Mrs. Toni Ross, 1974), 257. A fourth occupation could also be added as forestry work.

64. Sydney Hutcheson, PABC MS0030 (hereinafter Hutcheson ms. PABC).

65. See, for example, *Fernie Free Press*, 1 February 1901: "A Slav was injured in No. 2 mine on Tuesday"; 29 November 1902: "The deceased was a young man about 30. He had been in the camp only a few weeks."

66. For a discussion on Polish immigrants, see *Polish Settlers in Alberta: Reminiscences and Biographies*, ed. Joanna Matejko (Toronto: Polish Alliance Press, 1979), 19; Krystyna Lukasiewicz, "Polish Community in the Crowsnest Pass," *Alberta History* 36, no. 4 (Autumn 1988): 3. For general works, see Henry Radecki, *Ethnic Organizational Dynamics: The Polish Group in Canada* (Waterloo: Wilfrid Laurier University Press, 1979).

67. Ivy Haile, Fernie Interviews FDHS.

68. Testimony of Joe Fabin given to Alberta Coal Commission 1907, PAA 74.1,111 (hereinafter Coal Commission PAA).

69. *Crowsnest and Its People*, 481.

70. Harry Gate Interview, PAA 72.355, Taped and Written Interviews, 1972.

71. Hutcheson ms. PABC.

72. *Crowsnest and Its People*, 507.

73. *Vancouver Daily Province*, 27 May 1902, Scrapbook 1, Crowsnest Resources GA.

74. Ross, *Oh! The Coal Branch*, 234.

75. Andrew W. Harrell, *Accident History, Perceived Risk of Personal Injury, and Job Mobility as Factors influencing Occupational Accident Fatalism* (Edmonton: University of Alberta Centre for Experimental Sociology, 1985), 2.

76. Ibid., 5.

77. James F. Gavin and Robert E. Kelley. "The Psychological Climate and Reported Well-Being of Underground Miners: An Exploratory Study," *Human Relations* 31, no. 7 (July 1978): 568, 577. Despite this being a modern study, it is included here because of the basic correlations that can be drawn between the miners studied and the Crowsnest Pass before 1928. Gavin and Kelley interviewed only those miners who essentially carried on the same basic tasks as Crowsnest Pass miners, and at the same level of operation. The miners were underground workers in a small mine employing fewer than a thousand people, working in small crews on a rotating shift basis.

78. Frank Slide Visitor Centre, Hillcrest Mine Site Education Program, "Hear that Whistle Blow," 2001 (hereinafter "Hear that Whistle Blow").

79. Shaun Comish, *The Westray Tragedy: A Miners Story* (Halifax: Fernwood Publishing, 1993), 27, 34.

80. Green, "Calculated Risks," 182.

81. Christopher Monteressi, "Emotional Hijacking Versus Safe Behavior," Ph.D. diss., College of Engineering and Mineral Resources at West Virginia University, 1998, 4, 7.

82. District Inspector of Mines to Chief Inspector of Mines, 21 March 1917, file 204c, Reports PAA.

83. District Inspector of Mines to Chief Inspector of Mines, 3 March 1917, file 204c, Reports PAA.

84. James Crowder, District Inspector of Mines to John Stirling, Chief Inspector of Mines, 1919, file 88e, Reports PAA.

85 General Manager of Maple Leaf Coal Company, Bellevue to Chief Inspector of Mines, 8 August 1914, file 133i, Reports PAA.

86 Book Containing Minutes of Meetings of Directors of the Crowsnest Pass Coal Company Ltd. from 20 December 1905 to 7 August 1908, Crowsnest Resources GA.

87 Ibid., 28.

88 Inquest into the Death of Joseph Burns, 1926, file 204, Crowsnest Resources GA.

89 Directors Minutes, Crowsnest Resources GA.

90 *Fernie Free Press*, 2 August 1907.

91 Directors Minutes, Crowsnest Resources GA.

92 District Inspector of Mines to Chief Inspector of Mines, 28 September 1917, file 204c, Reports PAA.

93 Roy Lazarrotto Interview, Bellevue Mine, 11 November 2001.

94 *British Columbia Mines Annual Report 1908*, 242.

95 *British Columbia Mines Annual Report 1913*, 415; and *Annual Report 1914*, 507.

96 *British Columbia Mines Annual Report 1920*, 323. See also British Columbia, Attorney General – Inquisitions and Inquests, vol. 12, file 1921, PABC GR0431.

97 *Coleman Journal*, 24 May 1928.

98 Gino Bustao, Fernie Interviews FDHS.

99 *Crowsnest and Its People*, 819.

100 Monteressi, "Emotional Hijacking," 94.

101 Ibid., 29–30.

102 *Transcript of Evidence, Inquiry into the Fernie Coal Mine Explosion on 22 May, 1902, Volume 1*, Crowsnest Resources GA.

103 *Transcript of Evidence, Inquiry into the Explosion at Coal Creek Colliery 1915. In the Matter of the Public Inquiries Act, an Inquiry pursuant to a Commission dated 13 March 1915 directed by the Honourable the Lt. Gov. Of British Columbia Dr. John Stewart JP to the causes which led to an explosion in the Number B North Mine at Coal Creek on the 2nd day of January, 1915*, Crowsnest Resources GA.

104 Bellevue Mine Tour and Interview with Roy Lazarotto, 11 November 2001.

105 Crow's Nest Pass Coal Company, *Special Rules to be Observed by the Owners, Managers, Overmen, Master Mechanics, Firebosses & Workmen of the Collieries of the Crow's Nest Pass Coal Company, Limited* (n.p.: Crow's Nest Pass Coal Company, 1897).

106 Green, "Calculated Risks," 20–24.

107 *Transcript of Evidence, Inquiry into the Explosion at Coal Creek Colliery 1915*, Crowsnest Resources GA.

108 Gino Busato, Fernie Interviews FDHS.

109 James Taylor Adams, *Death in the Dark: A Collection of Factual Ballads of American Mine Disasters* (Virginia: Adams-Mullins Press, 1941), 4.

110 Inquest into the Death of Joseph Burns, 1926, Crowsnest Resources GA.

111 *Crowsnest and Its People*, 600.

112 Frank Anderson, *Tragedies of the Crowsnest Pass* (Surrey: Frontier Books, 1983), 49–52.

113 *Crowsnest and Its People*, 684.

114 *British Columbia Mines Annual Report 1917*, 313.

115 Kai T. Erikson, *Everything in its Path: Destruction of a Community in the Buffalo Creek Flood* (New York: Simon and Schuster, 1976), 165.

116 Judith Hoegg Ryan, *Coal in Our Blood: 200 Years of Coal Mining in Nova Scotia's Pictou County* (Halifax: Formac Publishing, 1992), 140.

117 *Crowsnest and Its People*, 476.

118 Ibid., 554.

119 Barry Paul Michrina, "Lives of Dignity, Acts of Emotion: Memories of Central Pennsylvania Bituminous Coal Miner Families," Ph.D. diss., California School of Professional Psychology, 1985, 175.

120 Ibid., 157.

121 Chrismas, *Alberta Miners*, 22.

122 Grace Dvorak, Fernie Interviews FDHS.

123 *Crowsnest and Its People*, 510.

124 Hazel Cerney, *The Silent Hills: the Kropinak family of Chapel Rock* (Blairmore: H.K. Cerney, 1992), 12.

125 *Crowsnest and Its People*, 554.

126 Brown ms. PABC, 213.

127 Joseph Gigliotti, Fernie Interviews FDHS.

128 Chrismas, *Alberta Miners*, 24.

129 Gino Remo, Fernie Interviews FDHS.

130 Elaine Leslau Silverman, *The Last Best West: Women on the Alberta Frontier 1880–1930* (Calgary: Fifth House, 1998), 76.

131 Alrik Tiberg, Oral History GA.

132 "Don't Go Below," in John O'Donnell, *And Now the Fields are Green: A Collection of Coal Mining Songs in Canada* (Sydney: University College of Cape Breton Press, 1992), 131.

133 Joseph Gigliotti, Dick and Dalia Guzzi, Fernie Interviews FDHS.

134 Chrismas, *Coal Dust Grins*, 74.

135 "The Miner's Doom," in Lloyd, *Come all Ye Bold Miners*, 202.

136 "Coal Mining Days," in O'Donnell, *And Now the Fields are Green*, 37.

137 Chrismas, *Alberta Miners*, 29.

138 Lynette Watchempino, "Dark as a Dungeon, Way Down in the Mine," *Suicide and Life-Threatening Behavior* 10, no. 4 (Winter 1980): 246.

139 Eric Margolis, "Western Coal Mining as a Way of Life: An Oral History of the Colorado Coal Miners to 1914," *Journal of the West* 24, no. 3 (July 1985): 55.

140 George Korson, *Minstrels of the Mine Patch* (Philadelphia: University of Pennsylvania Press, 1938), 94, 99.

141 Delphin A. Muise and Robert G. McIntosh, *Coal Mining in Canada: A Historical and Comparative Overview*, (Ottawa: National Museum of Science and Technology, 1996), 30.

142 Goodrich, *Miner's Freedom*, 22.

143 Priscilla Long, *Where the Sun Never Shines: A History of America's Bloody Coal Industry* (New York: Paragon House, 1989), 73.

144 Chrismas, *Alberta Miners*, 18.

145 Donovan, "Health and Safety," 97.

146 Chrismas, *Alberta Miners*, 24.

147 Ibid., 18

148 *Crowsnest and Its People*, 390.

149 Michrina, "Lives of Dignity," 119, 121.

150 Chrismas, *Alberta Miners*, 49.

151 Richard, *Westray Story* 1:vii.

152 *Crowsnest and It's People*, 554.

153 *Blairmore Enterprise*, 4 April 1929.

154 *Crowsnest and Its People*, 393.

155 "Down Deep in the Mine," in O'Donnell, *And Now the Fields are Green*, 55.

156 "Caledonia," ibid., 23.

157 *Coleman Journal*, 24 November 1927.

158 Janet Kennedy, *The McGillivray Creek Coal & Coke Co. and the International Coal and Coke Co., Coleman, Alberta: A Social History*, 1984, GA M7742 (hereinafter Kennedy ms. GA).

159 Ryan, *Coal in Our Blood*, 12.

160 Long, *Where the Sun Never Shines*, 34.

161 Makoto Takamiya, *Union Organization and Militancy: Conclusions from a Study of the United Mine Workers of America, 1940–1974* (Berlin: Verlag Anton Hain, 1978), 33. For a recent study that perhaps best exemplifies the humanness behind the miners' code, see Melissa Faye Greene, *Last Man Out: The Story of the Springhill Mine Disaster* (Toronto: Harcourt, 2003). Greene has taken the dispassionate psychological study of Rex Lucas regarding the 1958 Springhill mine explosion and assigned depth and personality to the men trapped for days in the mine.

162 *District Ledger*, 18 January 1908.

163 *Fernie Free Press*, 22 November 1902, 19 December 1903.

164 Attorney General – Inquisitions and Inquests, box 12, file 1466/05, PABC GR0429.

165 *British Columbia Mines Annual Report 1917*, 312.

166 George Wilkinson, Chief Inspector of Mines to Robert Strachan, Senior Inspector of Mines, Fernie, 12 March 1918, Bumps Correspondence FDHS.

167 F.H. Shepherd, *Blowouts of Gas, Morrissey British Columbia*, Bumps Correspondence FDHS.

168 Gino Busato, Fernie Interviews FDHS.

169 Robert Strachan, Senior Inspector of Mines to James Dickson, Chief Inspector of Mines, 21 April 1927, Bumps Correspondence FDHS.

Notes to Chapter 2

1 "The Boys of the Rescue Crew," in John C. O'Donnell, *And Now the Fields are Green: A Collection of Coal Mining Songs in Canada* (Sydney: University College of Cape Breton Press, 1992), 156–57.

2 Business letterhead of T.W. Davies on letter regarding compensation claim of Felix Vanduren, 7 September 1910, Attorney General Workman's Compensation Claims, box 1, file 4, Glenbow Archives 82.235 (hereinafter Compensation GA).

3 *Bellevue Times*, 28 February 1917.

4 Sworn statement of William McVey at the inquest into the death of William Archer, February 1917, Attorney General Inquest Files, file 966, PAA 67.172 (hereinafter Inquests PAA).

5 Report from the District Inspector to the Chief Inspector of Mines regarding the death of Steve Billesky, 3 March 1917, Mine Inspector's Reports, file 204c, PAA 77.237 (hereinafter Reports PAA).

6 Horace Harvey, *Report of the Inquiry into an Explosion which occurred on the 23rd Day of November, 1926 in a Coal Mine Operated by the McGillivray Creek Coal and Coke Company at Coleman Alberta*, 10 December 1927, 11.

7 Janet Kennedy, *The McGillivray Creek Coal & Coke Co. and the International Coal & Coke Co., Coleman, Alberta: A Social History*, Unpublished research paper, 1984, GA M7742 (hereinafter Kennedy ms. GA), 47.

8 *Coleman Journal*, 24 May 1928.

9 Report of a fatality at McGillivray Creek Coal and Coke Co. Ltd., Summary of Evidence, 6 August 1920, file 204d, Reports PAA.

10 District Inspector of Mines to the Chief Inspector of Mines with regard to a fatal accident that occurred 16 September 1914 in McGillivray Creek Mines, here called the Carbondale Mine, 24 September 1914, file 204b, Reports PAA.

11 Sworn statement of Albert T. Stewart, Mine Manager at the inquest of Albert Nycoma, file 934, Inquests PAA.

12 Duncan McDonald, "Mine Rescue and First Aid Work," *Bulletin of the Canadian Institute of Mining and Metallurgy* 115 (December 1921): 1051.

13 Kennedy ms. GA, 47.

14 R. Adams, *Eternal Prairie: Exploring Rural Cemeteries of the West* (Calgary: Fifth House, 1999), 81.

15 Frank B. Smith, "Coal Mining in the North-West Territories and its Probable Future," *Canadian Mining Review* 21, no. 4 (April 1902): 81.

16 *Annual Reports of the Mine Branch for the Province of Alberta* (hereinafter *Alberta Mines Annual Report*), 1905, 24.

17 *Coleman Journal*, 24 May 1928.

18 Letter from McGillivray Creek Coal and Coke Co. Ltd. to the Chief Inspector of Mines, 13 December 1918, file 204c, Reports PAA.

19 *Coleman Journal*, 24 May 1928.

20 Inquest into the death of William Dunn, Attorney General – Inquisitions and Inquests, vol. 11, file 1918, PABC GR0431.

21 Letters between the McGillivray Creek Coal and Coke Co. Ltd. and the District Inspector of Mines, 13 and 21 December 1918, file 204c, Reports PAA.

22 Letter from the District Inspector to the Chief Inspector regarding the death of John Bielec, who came to his death as

the result of injuries Friday, February 29, 1924, file 2041, Reports PAA.

23   Ibid. See also the Alberta Provincial Police Report on the death of John Bielec, 7 March 1924, box 34, file 1812, Inquests PAA, where the doctor's evidence at the inquest stated that he had received notice of a serious accident but that he waited at the hospital for the man to be sent to him instead of going to the mine.

24   Report from the Alberta Provincial Police on the Inquest for Robert Morgan, 5 September 1923, box 32, file 1793, Inquests PAA.

25   W.J. Dick, *Mine-Rescue Work in Canada* (Ottawa: Rolls L. Crain Co., 1912), 7, 13.

26   Donovan, "Health and Safety in Underground Coal Mining," 109–10.

27   Dick, *Mine-Rescue Work in Canada*, 17.

28   McDonald, "Mine Rescue and First Aid Work," 1051.

29   Dick, *Mine-Rescue Work in Canada*, 26.

30   Ibid., 35–56.

31   Julie Anne Green, "Calculated Risks: Worker, Owner and Government Attitudes Toward Safety in the Crow's Nest Pass Coal Mines, 1900–1915," Master's thesis, University of Calgary, 1990, Table 4, 195.

32   An Act to make Regulations with respect to Coal to Mines, *Statutes of the Province of Alberta Passed on the First Session of the First Legislative Assembly* (Edmonton: Kings Printer, 1906), 244, 250.

33   An Act Respecting Mines, *Revised Statutes of Alberta, 1922* (Edmonton: Kings Printer, 1922), section 86 (revised 1913), 2441.

34   Dick, *Mine-Rescue Work in Canada*, 25.

35   Book Containing Minutes of Meetings of the Directors of the Crowsnest Pass Coal Company from 7 August 1908 to 10 June 1921, Crowsnest Resources Ltd., vol. 9, 298, GA M1561 (hereinafter Directors Minutes, Crowsnest Resources GA).

36   "Memorandum of Experiences and Impressions in Connection with the Bellevue Explosion," prepared by Norman Fraser, file 87g, Reports PAA.

37   Ibid.

38   Ibid.

39   "My Account of the Bellevue Mine Explosion of December 10, 1910," by Andrew Matusky, file 87g, Reports PAA.

40   Fraser Memo, Reports PAA.

41   Frank B. Smith and Raoul Green, "The Frank Disaster" *The Canadian Mining Review* 22, no. 5 (May 1903): 102.

42   Frank Slide Visitor Centre, Hillcrest Mine Site Education Program "Hear that Whistle Blow," Point E, 2001 (hereinafter "Hear that Whistle Blow").

43   *Calgary Daily Herald*, 19 June 1914.

44   Saad, Michael, "Mining Disasters and Rescue Operations at Michel Before World War Two," *in The Forgotten Side of the Border: British Columbia's Elk Valley and Crowsnest Pass*, eds. Wayne Norton and Naomi Miller, (Kamloops: Plateau Press, 1998), 129.

45   James Ashworth, "Firedamp: Hydrogen, Methane, Ethane and Propane, Plus Air in British Columbia," *Monthly Bulletin of the Canadian Mining Institute* 94 (February 1920): 126.

46   *Fernie Free Press*, 9 January 1904.

47   William Hutchinson autobiographical manuscript, GA M558 (hereinafter Hutchinson ms. GA), 34–35.

48   Frank Anderson, *Tragedies of the Crowsnest Pass* (Surrey: Frontier Books, 1983), 59.

49   See Scrapbook no. 1, Newspaper Clippings 1902–1903, , Crowsnest Resources Ltd., box 64, file 583, GA M1561 (hereinafter Scrapbook 1, Crowsnest Resources GA); and the *Cranbrook Herald*, 29 May 1902, Crowsnest Pass Railway Route, http://www.crowsnest.bc.ca/fernie01.html (accessed 26 March 2004).

50   Anderson, *Tragedies of the Crowsnest Pass*, 65.

51   It is also entirely possible that these reports were due to the bias of the newspapers. However, the number of

these reports seems to give credence to this observation.

52. Duncan McDonald, Superintendent Lethbridge Mine Rescue Station to John T. Stirling Chief Inspector of Mines, 18 July 1914, file 40a, Reports PAA.

53. Frank Anderson, *Tragedies of the Crowsnest Pass*, 63.

54. Hutchinson ms. GA, 35.

55. Dick, *Mine-Rescue Work in Canada*.

56. Ibid, 37.

57. *Alberta Mines Annual Report 1912*, 119.

58. Dick, *Mine-Rescue Work in Canada*, 39.

59. Judith Hoegg Ryan, *Coal in Our Blood: 200 Years of Coal Mining In Nova Scotia's Pictou County* (Halifax: Formac Publishing, 1992), 32.

60. Duncan McDonald, Superintendent Lethbridge Mine Rescue Station to John T. Stirling Chief Inspector of Mines, 18 July 1914, file 40a, Reports PAA.

61. Dick, *Mine-Rescue in Canada*, 41.

62. John Kinnear, "Fred Alderson: The Hosmer Hero," in *The Forgotten Side of the Border: British Columbia's Elk Valley and Crowsnest Pass*, eds. Wayne Norton and Naomi Miller (Kamloops: Plateau Press, 1998), 179.

63. Ibid., 181–82.

64. *British Columbia Mines Annual Report 1915*, 323–24.

65. James Ashworth to John T. Stirling, Chief Inspector of Mines, 12 January 1915, file 40g, Reports PAA.

66. *Annual Report to the Minister of Mines, 1915*, 323–24.

67. Dick, *Mine Rescue in Canada*, 47-48

68. Ibid., 48.

69. James Ashworth, "Firedamp," 126–27.

70. "Hear that Whistle Blow," Point D, 2001.

71. Toledo Area Disaster Medical Assistance Team (TADMAT), "Blast Injuries" by Nick Olovos, http://mediccom.org/public/tadmat/training/ndms/Blast_Document.pdf (accessed 26 March 2004).

72. Harvey, *Report of an Enquiry into the Explosion*, 10.

73. "The Hillcrest Mine Disaster" as told to Stanley W. Frolick by his father, a survivor of the 1914 mine explosion, Hillcrest (Miscellaneous), PAA 75.198 (hereinafter Hillcrest misc. PAA), 34.

74. His name is variably written as True Witherby (*Toronto Globe*, 23 May 1902), Weatherby (*Rossland World*, 27 May 1902) or Trueman Weatherley (*Montreal Daily Star*, 14 June 1902).

75. *Montreal Daily Star*, 14 June 1902, Scrapbook 1, Crowsnest Resources GA.

76. *Rossland World*, 26 May 1902, Scrapbook 1, Crowsnest Resources GA.

77. Crowsnest Pass Historical Society, unpublished museum notes on Hillcrest Explosion, "Mine Ventilation."

78. Lars Weisaeth, "Psychological and Psychiatric Aspects of Technological Disasters," in *Individual and Community Responses to Trauma and Disaster: The Structure of Human Chaos*, eds. Robert J. Ursano, Brian G. McCaughey, and Carol S. Fullerton (Cambridge: Cambridge University Press, 1994), 81.

79. *Alberta Mines Annual Report 1912*, 120.

80. John T. Stirling, "Coal Mining in Alberta during 1914," *Coal Age* 7, no. 2 (9 January 1915): 62.

81. Green, "Calculated Risks," Table 3, 194.

82. See, for example, *British Columbia Mines Annual Report 1919*, 347.

83. See Green, "Calculated Risks," 64, and also ibid.

84. *Alberta Mines Annual Report 1916*, 68.

85. McDonald, "Mine Rescue and First Aid Work," 1051.

86. *British Columbia Mines Annual Report 1922*, 267.

87. See, for example, *District Ledger*, 14 May 1910; *Blairmore Enterprise*, 1 February 1912 and 28 March 1914; *Fernie Free Press*, 8 December 1926; and *British Columbia Mines Annual Report 1912*, 207. Green, in "Calculated Risks," speculates that once both mine rescue and first aid reached the levels of community sport through these

88 F.W. Gray, "Mine Rescue Work in Canada," *Canadian Mining Journal* (February 1913): 116.

89 Green, "Calculated Risks," 64.

90 Weisaeth, "Psychological and Psychiatric Aspects of Technological Disasters," 81–82.

91 *Blairmore Enterprise*, 20 July 1922.

92 *Blairmore Enterprise*, 16 February 1928.

93 *Coleman Journal*, 1 March 1928.

94 *Coleman Journal*, 24 May 1928.

95 *Coleman Miner*, 11 January 1911.

96 Duncan McDonald, Superintendent Lethbridge Mine Rescue Station to John T. Stirling Chief Inspector of Mines, 18 July 1914, file 40a, Reports PAA.

97 *Bellevue Times*, 18 December 1914.

98 Letter from the District Inspector to the Chief Inspector regarding the death of John Bielec, who came to his death as the result of injuries Friday, February 29, 1924, file 204I, Reports PAA.

99 Shaun Comish, *The Westray Tragedy: A Miner's Story* (Halifax: Fernwood Publishing, 1993), 43.

100 Harvey, *Report of the Inquiry into an Explosion*, 11–12.

101 Steve Stout, "Tragedy in November: The Cherry Mine Disaster," *Journal of the Illinois Historical Society* 72, no. 1 (February 1979): 57, 66.

102 "The Boys of the Rescue Crew," from O'Donnell, *And Now the Fields are Green*, 156–57.

103 *Alberta Mines Annual Report 1911*, 122.

104 Anderson, *Tragedies of the Crowsnest Pass*, 63.

105 Paul S. Fritz, "The Undertaking Trade in England: Its Origins and Early Development, 1660–1830," in *Eighteenth-Century Studies* 28, no. 2 (Winter 1994–95): 242.

106 Ibid., 241, 243–44.

107 Jani Scandura, "Deadly professions: Dracula, Undertakers and the Embalmed Corpse," *Victorian Studies* 40, no. 1 (Autumn 1996): 2.

108 Fritz, "The Undertaking Trade in England," 250.

109 Scandura, "Deadly Professions," 3, 9, 11.

110 Robert W. Habenstein and William M. Lamers, *The History of American Funeral Directing*, revised ed. (Milwaukee: Bulfin Printers, 1962), 225–27, 432–36.

111 For the Scott Brothers, see, for example, the *Pincher Creek Echo*, 6 September 1904; for R. Elliott see *District Ledger*, 18 January 1908; for Scott & Ross see *District Ledger*, 19 April 1905; and for A.E. Ferguson see ads in the 1914 *Fernie Free Press*.

112 See ads in the *District Ledger* for 18 April 1908, 20 June 1908, and 18 July 1908.

113 *District Ledger*, 7 June 1913.

114 *Coleman Journal*, 24 May 1928.

115 *Coleman Miner*, 18 June 1909.

116 Ibid., 2 July 1909.

117 T.W. Davies to Compensation Board, 7 September 1910, file 4, Compensation PAA.

118 Ibid., files 80 and 104.

119 Minute Book of the Coleman Local 2633, 1922–1924, 20 July 1924, GA M2239 (hereinafter Coleman Minutes GA).

120 *Blairmore Enterprise*, 7 July 1921.

121 *District Ledger*, 20 June 1908.

122 Storyboard on "The Transition Years" in the "Passing and Remembrance Exhibit," Sir Alexander Galt Museum and Archives, Lethbridge Alberta, June 2001.

123 Glennys Howarth, "Professionalising the Funeral Industry in England 1700–1960," in *The Changing Face of Death: Historical Accounts of Death and Disposal*, ed. Peter C. Jupp (London: MacMillan Press, 1997), 129.

124 The first official code of ethics was not adopted in Alberta until 1929, although it would probably have been a likely item of interest and discussion at the 1910 meeting Davies attended.

The British Code of Ethics would have been in force since at least 1905 with the establishment of the British Undertakers Association. See Harald Gunderson, *A History of Funeral Service in Alberta: Sixty Five Years of Serving Albertans, 1928–1993* (Calgary: Alberta Funeral Service Association, 1993), 23. See also Fritz, "The Undertaking Trade in England," 250.

125 C.F. Callaway, *The Art of Funeral Directing: A Practical Manual on Modern Funeral Directing Methods* (Chicago: Undertakers" Supply Co., 1928), Introduction.

126 Gunderson, *History of Funeral Service in Alberta*, 18. The Alberta Association did not lag far behind the British seeking to similarly organize themselves. In 1905, the British Undertakers organized as the central authority to govern and regulate the profession. See Fritz, "The Undertaking Trade in England," 250.

127 Gunderson, *History of Funeral Service in Alberta*, 27. The photograph shows T.W. Davies in the back row, third from the left. See also PAA photograph B7305, Convention of Alberta Undertakers in Edmonton, 1910.

128 Crowsnest Pass Historical Society, *Crowsnest and its People* (Coleman: Crowsnest Pass Historical Society, 1979), 488.

129 *Fernie Free Press*, 1 October 1920 (printed copy gratis of FDHS).

130 Corrine B. Lenfesty, "Choices for the Living, Honour for the Dead: A Century of Funeral and Memorial Practices in Lethbridge," Master's thesis, University of Lethbridge, 1998, 67. Lenfesty mentions this practice used by Lethbridge practitioners to "raise their profile."

131 *Blairmore Enterprise*, 27 January 1910.

132 Funeral Record Book, 1920–26, FDHS.

133 Howarth, "Professionalising the Funeral Industry in England," 129.

134 Robert J. Ursano and James E. McCarrol, "Exposure to Traumatic Death: The Nature of the Stressor," in *Individual and Community Reponses to Trauma and Disaster: The Structure of Human Chaos*, eds. Robert J. Ursano, Brian G. McCaughey, and Carol S. Fullerton (Cambridge: Cambridge University Press, 1994), 62–63.

135 Lorry Felske, "The Challenge Above Ground: Surface Facilities at Crowsnest Pass Mines Before the First World War," in *A World Apart: The Crowsnest Communities of Alberta and British Columbia*, eds. Wayne Norton and Tom Langford (Kamloops: Plateau Press, 2002), 164.

136 Thomas Lynch, *The Undertaking: Life Studies from the Dismal Trade* (New York: W.W. Norton & Co., 1997), 83.

137 *Calgary Daily Herald*, 20 June 1914.

138 *Coleman Bulletin*, 25 June 1914.

139 Ibid.

140 *Calgary Daily Herald*, 22 June 1914.

141 Interview with Mrs. J. Plante, 1972, PAA 72.355, Taped and written interviews, Item 2.

Notes to Chapter 3

1 "The Late Colliery Explosion at Patricroft, Wigan," in A.L. Lloyd, *Come all ye Bold Miners: Ballads and Songs of the Coalfields*, 2nd ed. (London: Lawrence and Wishart, 1978), 166–67.

2 For different views regarding disasters, see Anthony Oliver-Smith, "What is a Disaster?" in *The Angry Earth: Disaster in Anthropological Perspective*, eds. Anthony Oliver-Smith and Susanna M. Hoffman (New York: Routledge, 1999), 24; Wolf R. Dombrosky, "Again and Again: Is A Disaster What We Call A Disaster?" in *What is a Disaster? Perspectives on the Question*, ed. E.L. Quarantelli (London: Routledge, 1998), 21; Gary A. Kreps, "Disaster as Systemic Event and Social Catalyst," in *What is a Disaster?*, 33; and Boris N. Porfiriev, "Issues on the Definition and Delineation of Disasters and Disaster Areas," in *What is a Disaster?*, 60.

3. Philip A. Kalisch, "Death Down Below: Coal Mine Disasters in Three Illinois Counties, 1904–1962," *Journal of the Illinois State Historical Society* 65, no. 1 (Spring 1972): 5.

4. Kai T. Erikson, *Everything in Its Path: Destruction of Community in the Buffalo Creek Flood* (New York: Simon and Schuster, 1976), 155.

5. A.J.W. Taylor, *Disasters and Disaster Stress* (New York: AMS Press, 1989), 13.

6. Rex Lucas, *Men in Crisis: A Study of a Mine Disaster* (New York: Basic Books, 1969), 11.

7. Glin Bennett, *Beyond Endurance: Survival at the Extremes* (New York: St. Martin's Press, 1983), 38.

8. Kenneth Hewitt, "Excluded Perspectives in the Social Construction of Disaster," in *What is a Disaster?*, 77–78.

9. Oliver-Smith, "What is a Disaster," 23.

10. Kai Erickson argues that "death ... in the mines offers at least a thin layer of insulation, if only in the sense that one's nerves are braced in anticipation and one's imagination has had the chance to rehearse the possibilities." (Erickson, *Everything in Its Path*, 165.) However, no matter how braced an individual is for the death of a neighbour or workmate, the loss of 128 (Coal Creek, 1902) or 189 (Hillcrest, 1914) is still staggering.

11. Kenneth V. Iserson, *Grave Words: Notifying Survivors about Sudden, Unexpected Deaths* (Tucson: Galen Press, 1999), 4.

12. *Pincher Creek Echo*, 26 November 1908.

13. Crowsnest Pass Historical Society, *Crowsnest and its People* (Coleman: Crowsnest Pass Historical Society, 1979), 296.

14. Frank Slide Visitor Centre, Hillcrest Mine Site Education Program "Hear that Whistle Blow," Julia Makin's Story, 2001 (hereinafter "Hear That Whistle Blow").

15. *Crowsnest and its People*, 574.

16. *Fernie Free Press*, 24 May 1902.

17. *Crowsnest and its People*, 906.

18. Janet Kennedy, *The McGillivray Creek Coal & Coke Co. and the International Coal & Coke Co., Coleman, Alberta: A Social History*, Unpublished research paper, 1984, 55, Glenbow Archives M7742 (hereinafter Kennedy ms. GA).

19. Judith Hoegg Ryan, *Coal in Our Blood: 200 Years of Coal Mining in Nova Scotia's Pictou County* (Halifax: Formac Publishing, 1992), 14.

20. Iserson, *Grave Words*, 5.

21. Jack McAndrew commentary, *Blood on the Coal* [video recording] (Toronto: CBC International Sales, 1999) (hereinafter *Blood on the Coal* video).

22. Iserson, *Grave Words*, 13

23. *Daily News of Nelson*, 25 May 1902 and 28 May 1902, Scrapbook 1, Newspaper Clippings 1902–1903, box 64, file 583, Crowsnest Resources, GA M1561 (hereinafter Crowsnest Resources GA).

24. Iserson, *Grave Words*, 249.

25. *Fernie Free Press*, 6 April 1917.

26. Thomas Lynch, *The Undertaking: Life Studies from the Dismal Trade* (New York: W.W. Norton & Co., 1997), 5.

27. See Frank Anderson, *Tragedies of the Crowsnest Pass* (Surrey: Heritage House Publishing, 1983), 53–54; "Hear that Whistle Blow," Point D, 2001.

28. *Fernie Free Press*, 18 April 1903.

29. *Toronto Globe*, 26 May 1902, Scrapbook 1, Crowsnest Resources GA.

30. Anderson, *Tragedies of the Crowsnest Pass*, 54.

31. "Report of Accident from Bellevue Mine Manager to Chief Inspector of Mines 23 March 1929," file 87f, Mine Inspector's Reports, PAA 77.237 (hereinafter Reports PAA).

32. "List of Injuries to Hillcrest Dead," file 40a, Reports PAA.

33. Attorney General – Inquisitions and Inquests, vol. 10, file 1915, PABC GR0431.

34. Bert Wentworth and Harris Hawthorne Wilder, *Personal Identification: Methods for the Identification of Individuals, Living or Dead* (Chicago: T.G. Cooke, 1932), 58.

35 "The Wallsend Explosion," in A.L. Lloyd, *Come All Ye Bold Miners: Ballads and Songs of the Coalfields* 2nd ed. (London: Lawrence and Wishart, 1978), 155.

36 R.A.H. Morrow, *Story of the Springhill Disaster: Comprising a Full and Authentic Account of the Great Coal Mining Explosion at Springhill Mines, Nova Scotia, February 21st, 1891* (Saint John: R.A.H. Morrow, 1891), 69.

37 Priscilla Long, *Where the Sun Never Shines: A History of America's Bloody Coal Industry* (New York: Paragon House, 1989), 44.

38 Sydney Hutcheson, *Depression Stories* (Vancouver: New Star Books, 1976), 14–15.

39 *Rossland World*, 28 May 1902, Scrapbook 1, Crowsnest Resources GA.

40 James Crowder, District Inspector of Mines to John Stirling, Chief Inspector of Mines, 5 December 1922 (this correspondence describes the initial accident and identification of the bones as belonging to Machin), file 88f, Reports PAA. James Crowder, District Inspector of Mines to John Stirling Chief Inspector of Mines, 2 March 1923 (this correspondence details the finding of Machin's body: "how he got into the river is unknown"), file 88g, Reports PAA.

41 "Transcript of Evidence from Inquest into Fernie Coal Mine Explosion on May 22nd 1902," file 213, Crowsnest Resources GA. Testimony of Henry Manley from the Engineering Department using blueprints to indicate where men were found at the time of their deaths.

42 *Fernie Free Press*, 18 April 1903.

43 Anderson, *Tragedies of the Crowsnest Pass*, 64.

44 Long, *Where the Sun Never Shines*, 67.

45 Horace Harvey, Report of the Inquiry into an Explosion which occurred on the 23rd Day of November, 1926 in a Coal Mine Operated by the McGillivray Creek Coal and Coke Company at Coleman Alberta, 10 December 1927, 11.

46 Anderson, *Tragedies of the Crowsnest Pass*, 65.

47 Inspector C. Junget, "Report on Mine Disaster at Hillcrest, Alta, Bellevue, Alta., 24 June 1914," attached as Appendix O to Crowsnest Pass Historical Society, unpublished museum notes on Hillcrest Explosion, copied August 2001.

48 *Calgary Daily Herald*, 11 December 1910.

49 "Hear that Whistle Blow," Julia Makin's Story, 2001.

50 Peter E. Hodgkinson and Michael Stewart, *Coping with Catastrophe: A Handbook of Post-Disaster Psychosocial Aftercare* 2nd ed. (London: Routledge, 1998), 35.

51 Hutcheson, *Depression Stories*, 14.

52 Crowsnest Pass Historical Society, unpublished museum notes on Hillcrest Explosion, copied August 2001.

53 *Bellevue Times*, 27 May 1902, Scrapbook 1, Crowsnest Resources GA.

54 Hodgkinson, *Coping with Disaster*, 36.

55 Melissa Faye Greene, *Last Man Out: The Story of the Springhill Mine Disaster* (Toronto: Harcourt, Inc., 2003), 244.

56 Lynch, *The Undertaking*, 20–21.

57 Erich Lindemann, "Symptomatology and Management of Acute Grief," *American Journal of Psychiatry* 101 (1944): 142–43.

58 *Bellevue Times*, 26 June 1914.

59 *Crowsnest and its People*, 770.

60 *Bellevue Times*, 26 June 1914.

61 "The frenzied crowd poured from their houses seeking information," *Daily News of Nelson*, 25 May 1902, Scrapbook 1, Crowsnest Resources GA; see also *Calgary Daily Herald*, 25 June 1914.

62 *Coleman Journal*, 25 November 1926.

63 Erikson, *Everything in its Path*, 58.

64 *Mail and Empire*, 26 May 1902, Scrapbook 1, Crowsnest Resources GA.

65 *Bellevue Times*, 19 June 1914.

66 *Crowsnest and Its People*, 510.

67 Lindemann, "Symptomatology and Management of Acute Grief," 146.

68 George Korson, *Minstrels of the Mine Patch* (Philadelphia: University of Pennsylvania Press, 1938), 184.

69 James Taylor Adams, *Death in the Dark: A Collection of Factual Ballads of American Mine Disasters* (n.p.: Adams-Mullins Press, 1941), 29.

70 For accounts of trapped miners, see Korson, *Minstrels of the Mine Patch*, 183; *Blood on the Coal* video; "Account of Cherry mine fire," *Toronto Globe*, 22 November 1909, Scrapbook no. 2, Newspaper Clippings 1903–1910, box 65, file 584, GA M1561 (hereinafter Scrapbook 2, Crowsnest Resources GA).

71 "Account of Cherry mine fire," *Toronto Globe*, 22 November 1909, Scrapbook 2, Crowsnest Resources GA.

72 Ibid.

73 Andreas Ploeger, "A 10-Year Follow up of Miners Trapped for 2 Weeks under Threatening Circumstances," *Stress and Anxiety* 4 (1977): 23.

74 For accounts of the Springhill mine disasters, see Roger David Brown, *Blood on the Coal: The Story of the Springhill Mining Disasters* (Windsor: Lancelot Press, 1976), 44–54; *Blood on the Coal* video; Lucas, *Men in Crisis*; and Greene, *Last Man Out*.

75 Lindemann, "Symptomatology and Management of Acute Grief," 142.

76 Margie Kempt commentary, *Blood on the Coal* video. The 1958 disaster was the second time Gorley Kempt had been trapped underground. His previous experience occurred two years earlier in the 1956 Springhill explosion. Kempt commented when he made it out alive after 1958: "Thank God I am alive. Twice is enough. I will never go down again." Brown, *Blood on the Coal*, 47.

77 Greene, *Last Man Out*, 20.

78 *Crowsnest and Its People*, 333.

79 J. William Kerr, *Frank Slide* (Priddis: Barker Publishing, 1990), 15.

80 Frank Anderson, *Tragedies of the Crowsnest Pass*, 49–52.

81 *District Ledger*, 27 June 1914.

82 Marvin R. Hershiser and E.L. Quarentelli, "The Handling of the Dead in a Disaster," *Omega* 7, no. 3 (1976): 202.

83 Grace Arbuckle Dvork, "Childhood Remembered: A Coal Creek Memoir," in *The Forgotten Side of the Border: British Columbia's Elk Valley and Crowsnest Pass*, eds. Wayne Norton and Naomi Miller (Kamloops: Plateau Press, 1998), 193.

84 Charles E. Fritz and J.H. Mathewson, *Convergence Behavior in Disasters: A Problem in Social Control*, Disaster Study No. 9 (Washington: National Research Council, 1957), iii.

85 Kerr, *Frank Slide*, 24–25.

86 Lars Weisaeth, "Psychological and Psychiatric Aspects of Technological Disasters," in *Individual and Community Responses to Trauma and Disaster: The Structure of Human Chaos*, eds. Robert J. Ursano, Brian G. McCaughey, and Carol S. Fullerton (Cambridge: Cambridge University Press, 1994), 86.

87 See *Coleman Journal*, 25 November 1926: "Several hundred people gathered to secure first hand knowledge of how it fared with their loved ones." Also *Calgary Daily Herald*, 10 December 1910: "When the explosion in the Bellevue mine ... occurred last night ... there was such a rush of the several hundred employees and friends and relatives."

88 *Crowsnest and Its People*, 59.

89 *Bellevue Times*, 16 December 1910.

90 *Crowsnest and Its People*, 66.

91 Sydney Hutcheson, PABC MS0030 (hereinafter Hutcheson ms. PABC).

92 William Hutchinson autobiographical manuscript, GA M558 (hereinafter Hutchinson ms. GA), 35.

93 *Crowsnest and Its People*, 798.

94 "Hear that Whistle Blow," Julia Makin's Story, 2001.

95 *Crowsnest and Its People*, 483.

96 Interview with Harry Gate, Coleman, 1972, PAA 72.355, Taped and written interviews, Item 4.

97 For a full explanation of the 1902 conflict, see Andrew Yarmie, "Community and Conflict: The Impact of the 1902 Explosion at Coal Creek," in *The Forgotten Side of the Border: British Columbia's Elk Valley and Crowsnest Pass*, eds. Wayne Norton and Naomi Miller (Kamloops: Plateau Press, 1998), 195–205.

98 *Toronto Globe*, 26 May 1902, Scrapbook 1, Crowsnest Resources GA.

99 *Fernie Free Press*, 24 May 1902.

100 E.L. Quarantelli and Russell R. Dynes, "Community Conflict: Its Absence and Its Presence in Natural Disasters," *Mass Emergencies* 1 (1976): 140.

101 *Calgary Daily Herald*, 11 December 1910.

102 *Bellevue Times*, 26 June 1914.

103 Chief Inspector of Mines to Proprietor of Hillcrest Hotel, 16 July 1914, file 40c, Reports PAA.

104 For discussions on ethnic mutual aid societies, see Howard Palmer, *Land of the Second Chance: A History of Ethnic Groups in Southern Alberta* (Lethbridge: Lethbridge Herald, 1972); Donald Avery, *"Dangerous Foreigners:" European Immigrant Workers and Labour Radicalism in Canada, 1896–1932* (Toronto: McClelland and Stewart, 1979); Patricia K. Wood, *Nationalism from the Margins: Italians in Alberta and British Columbia* (Montreal: McGill-Queen's University Press, 2002); John Norris, *Strangers Entertained: A History of the Ethnic Groups of British Columbia* (Vancouver: Evergreen Press, 1971); and Interview with Angelo Toppano, Coleman, 1972, PAA 72.355, Taped and written interviews, Item 3.

105 For general works on fraternal societies see Clifford Putney, "Service over Secrecy: How Lodge-Style Fraternalism yielded to Popularity with Men's Service Clubs," *Journal of Popular Culture* 27, no. 1 (1993): 179–90; *The Cyclopaedia of Fraternities*, ed. Albert C. Stevens (New York: E.B. Treat and Co., 1907); Mary Ann Clawson, "Fraternal Orders and Class Formation in the Nineteenth Century United States," *Comparative Studies in Society and History* 27, no. 4 (October 1985): 672–95; and David T. Beito, "Thy Brother's Keeper: The Mutual Aid Tradition of American Fraternal Orders," *Policy Review* 70 (Fall 1994).

106 *Crowsnest and Its People*, 569.

107 Independent Order of Oddfellows, Mount Fernie Lodge No. 47, *Constitution, By-Laws and Rules of Order of Mount Fernie Lodge No. 47, IOOF under Jurisdiction of the Grand Lodge of British Columbia, Fernie, BC, Instituted February 24, 1902*, 14.

108 *Bellevue Times*, 16 June 1911.

109 *Crowsnest and Its People*, 313.

110 Interview with Angelo Toppano, Coleman, 1972, PAA 72.355, Taped and written interviews, Item 3.

111 *Crowsnest and Its People*, 325.

112 Independent Order of Oddfellows, *Constitution*, 16.

113 *Funeral Record Book, 1920–1926*, Fernie and District Historical Society.

114 *Report of Enquiry as to Funeral Costs in the Province of Alberta, February 11, 1932*, Alex Ross to the Honorable Lieutenant Governor-General of Alberta, Sessional Paper no. 72, 2. This is borne out by a district by district account of the amount charged for funerals, taken from *Register regarding Investigation of Funeral Costs*, PAA 89.463, box 2. Although funeral costs are not available for the Alberta towns of the Crowsnest Pass area, funerals in Drumheller ranged from $15 to $450 with over half in the $100 range. Funerals in Lloydminster ranged from $75 to $261.65 with over half in the $100 range. Funerals in Macleod ranged from $14 to $575. The *Funeral Record Book, 1920–1926* for Fernie has ranges for funerals with a high of $301 to less than $100. The average appears to be somewhere in the $120–$150 range.

115 *Memoirs of Polish Immigrants in Canada*, ed. Benedykt Heydenkorn (Toronto: Canadian-Polish Research Institute, 1979), 42. See also GA M218, Chuchla Family.

116 Krystyna Lukasiewicz, "Polish Community in the Crowsnest Pass,"

117 *Crowsnest and Its People*, 314.

118 For more information on unions, see Frank Paul Karas, "Labour and Coal in the Crowsnest Pass: 1925–1935," Master's thesis, University of Calgary, 1972; Warren Caragata, "The Labour Movement in Alberta: An Untold Story," in *Essays on the Political Economy of Alberta*, ed. David Leadbetter (Toronto: New Hogtown Press, 1984); Makoto Takamiya, *Union Organization and Militancy: Conclusions from a Study of the United Mine Workers of America, 1940–1974* (Berlin: Verlag Anton Hain, 1978); Elizabeth Levy and Tad Richards, *Struggle and Lose, Struggle and Win: The United Mine Workers* (New York: Four Winds Press, 1977); Allen Seager, "Socialists and Workers: The Western Canadian Coal Miners, 1900–1921," *Labor/Le Travail* 16 (Fall 1985): 23–59; and Wesley Morgan, "The One Big Union and the Crowsnest Pass," in *A World Apart: The Crowsnest Communities of Alberta and British Columbia*, eds. Wayne Norton and Tom Langford (Kamloops: Plateau Press, 2002).

119 United Mine Workers of America Constitution and Bylaws 1915–1955, box 15, file 138, GA M2239 (hereinafter United Mine Workers GA).

120 Strike action could also result in the betterment of safety. The Michel strike of 1911 resulted in the installation of a phone line in the mine to assist with accidents or equipment damage. The 1918 strike in Fernie and Michel resulted in the single-shift clause to mitigate what was recognized as dangerous conditions in the mine during a double-shift system. See Department of Labour, Economics & Research Branch – Strikes & Lockouts, PABC GR1695 (hereinafter Strikes & Lockouts PABC), Reels B07127 and B07134.

121 Both Wesley Morgan, "The One Big Union and the Crowsnest Pass," 113, and Brown ms. PABC, 193 make this assertion.

122 Karas, "Labour and Coal," 32.

Alberta History 36, no. 4 (Autumn 1988): 5.

123 *British Columbia Mines Annual Report 1903*, 227.

124 *Fernie Free Press*, 28 June 1901.

125 Report from Chief Constable's Office in Nelson, British Columbia, January 1902, Attorney General – Inquisitions and Inquests, box 8, PABC GR0429.

126 *Calgary Herald*, March 1910.

127 See a series of articles in the *Fernie Free Press* between 21 February 1903 and 4 April 1903.

128 *District Ledger*, 17 June 1911.

129 *Labour Gazette*, January 1932, 92–103.

130 See, for example, the strikes at Frank in 1911, Blairmore in 1920, and Fernie in 1920, Reels B07127, B07144, B07147, Strikes & Lockouts PABC.

131 *Fernie Free Press*, 16 February 1912; *District Ledger*, 10 February 1912.

132 Reel B07151, Strikes & Lockouts PABC.

133 Levy and Richards, 28.

134 See, for example, *District Ledger*, 17 February 1912.

135 Coleman Local 2633 Minutes 1903–1909, GA M2239 (hereinafter Local 2633 GA).

136 File 183, Crowsnest Resources GA.

137 *Funeral Records Book, 1920–1926*, FDHS.

138 Local 2633 GA.

139 *Coleman Miner*, 17 July 1908.

140 Report from the Alberta Provincial Police to the Attorney General, 23 July 1926, Attorney General Inquest Files, box 257, file 17, PAA 68.261 (hereinafter Inquests PAA).

141 Local 2633 GA.

142 Testimony given to the Alberta Coal Commission, file 321, PAA 74.1 (hereinafter Coal Commission PAA).

143 Alberta first proclaimed a Compensation Act in 1908: "An Act with Respect to Workmen for Injuries Suffered in the Course of Their Employment," *Statutes of the Province of Alberta Passed on the Third Session of the Legislative Assembly* (Edmonton: King's

Printer, 1908). Despite this, given the correspondence in the Compensation claim files, receiving money from the government for deaths suffered in mining accidents was neither easy nor assured. See Compensation PAA.

144 General Manager of the Crowsnest Pass Coal Company to the Attorney General, 31 October 1905, British Columbia Attorney General, box 12, file 3150/05, PABC GR0429 (hereinafter Inquests PABC).

145 File 321, Coal Commission PAA.

146 See ad for "Miner's Special Policies" in the *Fernie Free Press*, 22 November 1902.

147 Probate PABC.

148 Register of Marriages, box 30, file 3, Anglican Church of Canada, Calgary Diocese (hereinafter ACCCD). Numbers of women remarrying are unavailable for the Catholic Diocese.

149 Box 1, file 1, Compensation PAA.

150 *Crowsnest Pass Historical Driving Tour: Blairmore*, 12.

151 *Backtracking with the Fernie Historical Association*, 71; Rev. F.E. Runnalls, *It's God's Country: A Review of the United Church and Its Founding Partners, the Congregational, Methodist and Presbyterian Churches in British Columbia* (n.p., 1974), 158–59.

152 *Coleman's Fiftieth Anniversary Booklet* (Lethbridge: Lethbridge Herald, 1953), 38.

153 Robinson correspondence 1906, box 20, file 25, ACCCD.

154 Peoples Warden to Archdeacon Hayes, June 1926, box 6, file 32, ACCCD.

155 Coleman Service Register, 1921–1949, box 31, file 7, ACCCD.

156 For a thorough study of the problems faced by the Catholic Church in the Diocese of Calgary (which encompassed all of southern Alberta), see Jo-Ann Munn Gafuik, "Bishops, Priests and Immigrants: The Roman Catholic Diocese of Calgary and the Immigrant Question, 1912–1967," Masters thesis, University of Calgary, 1996.

157 *Crowsnest and Its People*, 292.

158 *History of St. Anne's Parish, Blairmore – Jubilee Book 1910–1985*, 3.

159 Notes from Father Beaton on the founding of the Bellevue Parish, file 55.1345, Roman Catholic Diocese Archives (hereinafter RCDA).

160 Father Cosman to Bishop McNally, 2 July 1917, RCDA.

161 Father Kreciszewski to Bishop McNally, 7 July 1922, RCDA.

162 Norman Knowles, "A Manly, Commonsense Religion': Revivalism and the Kootenay Campaign of 1909 in the Crowsnest Pass," in *A World Apart: The Crowsnest Communities of Alberta and British Columbia*, eds. Wayne Norton and Tom Langford (Kamloops: Plateau Press, 2002), 3.

163 John Bielec to Father McNally, 18 October 1916, file 63.1441, RCDA.

164 C.A. McDonald to Bishop McNally, 15 December 1923, file 55.1345, RCDA.

165 M.A. Harrington, *Twenty-Fifth Anniversary, St. Anne's Parish, Blairmore, Alberta* (Blairmore: Enterprise Job Print, 1935), 15.

166 Files 53.145, 56.1361, 63.1441, RCDA.

167 *Crowsnest and Its People*, 293. See also M.B. Venini Byrne, *From the Buffalo to the Cross: A History of the Roman Catholic Diocese of Calgary* (Calgary: Calgary Archives and Historical Publishers, 1973), 226.

168 Quoted in Munn Gafuik, "Bishops, Priests and Immigrants," 98.

169 Runnalls, *It's God's Country*, 159.

170 Hershiser and Quarentelli, "The Handling of the Dead in a Disaster," 196.

171 Erikson, *Everything in its Path*, 169.

172 *Bellevue Times*, 23 December 1910.

173 *Crowsnest and Its People*, 510–11.

174 Harvey, *Report of the Inquiry into an Explosion*, 8–9.

175 John Kinnear, "Fred Alderson: The Hosmer Hero," in *The Forgotten Side of the Border: British Columbia's Elk Valley and Crowsnest Pass*, eds. Wayne Norton

and Naomi Miller (Kamloops: Plateau Press, 1998), 183.

176 "A Miner's Life," from John C. O'Donnell, *"And Now the Fields are Green": A Collection of Coal Mining Songs in Canada* (Sydney: University College of Cape Breton Press, 1992), 120.

177 "The Man with a Torch in His Cap," from O'Donnell, *"And Now the Fields are Green,"* 124.

178 *Alberta Mines Annual Report 1911*, 122.

179 Letter from James Ashworth to the editor of the Canadian Mining Journal entitled "Exclusions from Falls of Roofs in Mines," file 87a, Reports PAA.

180 *Calgary Daily Herald*, 10 December 1910.

181 *Coleman Miner*, 23 December 1910.

182 Kinnear, "Fred Alderson," 183.

183 *Bellevue Times*, 23 December 1910.

184 *Bellevue Times*, 24 March 1911.

185 Arthur L. Sifton, Minister of Public Works to His Honour the Lieutenant Governor in Council, 11 March 1911, Legislative Assembly, file 38, PAA 70.414.

186 Brown ms. PABC, 218.

187 *Bellevue Times*, 24 March 1911.

188 *District Ledger*, 14 May 1910.

189 See John Benson, "Charity's Pitfalls: The Senghenydd Disaster," *History Today* 43 (November 1993): 7–9. Benson details the 1913 coal mine disaster at Senghenydd, Wales and the desire of both management and government to keep up the appeals for relief funds. Benson quotes one of the owners of the Swaithe Main colliery: "I am convinced that if we do not subscribe the whole burden will fall upon us, whereas if we do subscribe, I think our efforts will be supplemented by the liberality of the public."

190 *Toronto Globe*, 27 May 1902, Scrapbook 1, Crowsnest Resources GA.

191 *Nelson News*, 30 May 1902, Scrapbook Number 1, Crowsnest Resources GA.

192 *Daily News*, 30 May 1902, Scrapbook Number 1, Crowsnest Resources GA.

193 *Fernie Free Press*, 14 June 1902.

194 Michael C. Kearl, *Endings: A Sociology of Death and Dying* (New York: Oxford University Press, 1989), 478.

195 Lucas, *Men in Crisis*, 9.

196 Robert W. Habenstein and William M. Lamers, *The History of American Funeral Directing* (Milwaukee: Bulfin Printers, 1962), 354.

197 Lynch, *The Undertaking*, 109.

198 *Coleman Journal*, 7 January 1926.

199 *Blairmore Enterprise*, 22 May 1914.

200 Douglas J. Davies, *Death, Ritual and Belief: The Rhetoric of Funerary Rites* (London: Cassell Publishers, 1997), 57.

201 Daniel David Cowell, "Funerals, Family, and Forefathers: A View of Italian-American Funeral Practices," *Omega* 16, no. 1 (1985–86): 71.

202 Sula Benet, *Song, Dance and Customs of Peasant Poland* (London: Dennis Dobson, 1951), 240–42.

203 *Crowsnest and Its People*, 848.

204 "The Horse Drawn Funeral in Early America," *The Driving Digest Magazine* 105 (1998): 10.

205 Bruce Ramsay, *100 Years of Coal Mining: The Elk River Valley* (Sparwood: Ramsay Publications, 1997), 96

206 *Bellevue Times*, 16 December 1910.

207 *Fernie Free Presss*, 1 April 1904.

208 *Bellevue Times*, 27 September 1912.

209 *Fernie District Ledger*, 19 September 1918.

210 *Crowsnest and Its People*, 320.

211 Lynch, *The Undertaking*, 81.

212 Several authors make note of the social nature of funerals. Habenstein notes that "funeral processions are solemn social events" (Habenstein and Lamers, *History of American Funeral Directing*, 354), while Sloane charts the progress of funerals from purely religious occasions to ones of social significance (David Charles Sloane, *The Last Great Necessity: Cemeteries in American History* (Baltimore: Johns Hopkins University Press, 1991), 26). Thomas Lynch

notes especially that "funerals are the emotionally potent and spiritually stimulating intersection of the living and the dead." (Lynch, *The Undertaking*, 81.)

213 *Pincher Creek Echo*, 24 August 1923 and 31 August 1923.

214 Joseph J. Edgette, "The Epitaph and Personality Revelation," in *Cemeteries and Gravemarkers: Voices of American Culture*, ed. Richard E. Meyer (Ann Arbor: UMI Research Press, 1989), 87. Edgette remarks that "memorials erected to honor the dead serve an important and often complex function within the society which creates them."

215 Nancy Millar, *Remember Me as You Pass By: Stories from Prairie Graveyards* (Calgary: Glenbow, 1994), 157. Millar writes: "We will not be denied our individuality."

216 Conversation with Fernie Cemetery grounds staff, September 2001.

217 *Bellevue Times*, 23 December 1910.

218 Probate PABC.

219 Nancy Millar, *Once Upon a Tomb: Stories from Canadian Graveyards* (Calgary: Fifth House, 1997), 229–30. Millar makes note of grave markers and monuments found in New Glasgow, Westville, and Springhill.

220 Duncan Emrich, "Songs of the Western Miners," *California Folklore Quarterly* I (1942): 221.

221 O'Donnell, *"And Now the Fields are Green"*, 187, 188, 198.

222 Ibid., 39.

223 "Hillcrest Mine," by James Keelaghan, in *Small Rebellions* [sound recording], 1990.

224 Conversation with Monica Field and Diane Peterson, Frank Slide Interpretative Centre, June 2001.

225 Sheila Adams, "Women, Death and *In Memoriam* Notices in a Local British Newspaper," in *The Unknown Country: Death in Australia, Britain and the USA*, eds. Kathy Charmaz, Glennys Howarth, Allan Kellehear (New York: St. Martin's Press, 1997), 98.

226 *Pincher Creek Echo*, 29 June 1923; *Coleman Miner*, 2 April 1909; *Crowsnest and Its People*, 320. See also *Coleman Journal*, 26 June 1926, where the Oddfellows and the Rebekahs marched to the cemetery to decorate the graves of the dead.

227 *Bellevue Times*, 15 December 1911.

228 *Blairmore Enterprise*, 7 July 1916. See also printed copy of the memorial address "In Memory of Hillcrest Disaster June 19 1914," PAA 86.67, Speech by Edward Keith on 1914 Hillcrest Disaster; *Blairmore Enterprise*, 18 June 1925.

229 *Fernie Free Press*, 23 May 1903; 20 May 1910; 4 July 1913.

230 Since 1899, the community of Mt. Olive, Illinois has set aside 12 October to commemorate the dead in the Union Miners Cemetery. The community celebrates the "acts of heroism in organizing [union] campaigns, the everyday tragedies involved in the coal miner's job and the vicious infighting among the miners themselves." See John H. Keiser, "The Union Miners Cemetery at Mt. Olive, Illinois: A Spirit-Thread of History," *Journal of the Illinois State Historical Society* 62, no. 3 (Autumn 1969): 232. Mt. Olive, however, has a distinguished clientele residing in the cemetery given the flamboyant history of unionizing in the area. Mary Harris Jones, or "Mother Jones," is buried at Mt. Olive. Another example of continually held anniversary memorials is the annual services held to commemorate the victims of the Millfield, Ohio explosion that killed 82 men in 1930. See Daniel David Crowell, "Death Underground: The Millfield Mining Tragedy," *Timeline* 14, no. 5 (September/October 1997): 42–56.

231 E-mail from Cathy Pisony, Frank Slide Interpretative Centre, 7 November 2001. The community has long felt the need for a monument. At one time the EcoMuseum Trust raised money to have a commemorative plaque posted by the mass graves. However, the gravesite at Hillcrest is a Provincial Historic Site and cannot be changed in any way. The Crowsnest Pass experience in attempting

over the years to erect large monuments to the mining dead is not unique. Another example of later generations seeking to create a monument is given by Father Briggs at Monongah, West Virginia, who "devoted more than half his life to memorializing those who died in the disaster." See Barbara Smith, "I Know Them All: Monongah's Faithful Father Briggs," *Goldenseal* 25, no. 4 (Winter 1999): 21–24.

232 Lucas, *Men in Crisis*, 40–41.

INDEX

Alderson, Fred, 49, 109–10, 131, 134
Archer, William, 34–35
Ash, Edgar, xxiii, 101
Bardsley, Thomas, 122, 124
Bellevue Mine explosion (1910), 41–43, 85, 95, 108, 112, 137
Bielec, John, 38, 55, 106
Billesky, Steve, 17, 35
blackdamp, 6, 90
bumps and blowouts, xvi, 8, 31–32
Burns, Joseph, 17, 22, 136
Busato, Gino, 12, 19, 21,
camaraderie, 25, 28–29, 56
Canadian-American Coal and Coke Company, 2
cemeteries. *See* grave markers
Chuchla, Walter, 98
churches, 103–7
Clarke, William, 131, 134
Coal Creek explosion (1902), xiii, xv, 4–5, 15, 44, 45, 79, 82, 84, 90, 95, 112, 113, 118–19
Coal Creek explosion (1917), xiii, 7, 79, 83, 86, 124
coal dust, 5
coalification, 5
Coats, Charles, 124, 130
Coleman Brotherly Aid Society. *See* mutual aid societies
compensation, xviii, 12, 60–61, 102, 103
convergence behaviour. *See* disasters
Crowsnest Pass, 165n4
    description of coal basin, 2
    geographical location, x, xiv
    population of, xv, 3
Crows Nest Pass Coal Company, 2, 40, 111, 167n4
Davies, T.W. *See* undertakers
dead, identification of, 79, 82–86
disasters. *See also* explosions

convergence behaviour, 93–96
    definitions of, 76–77, 177n2
    psychology of, xxiii, 77, 91–92, 166n30, 172n161
    statistics, 147, 151–52
    time phases of, 52, 54
Draeger device. *See* mine rescue, equipment
Edwin, Samuel, 124, 127
Elick, Charles, 22, 92
ethnicity, xiv–xv, 45, 95, 96, 100–101, 105, 106, 144, 148–50
Evans, Evan, xxiii, 49–50, 131–32
explosions
    description and force of, 5–6, 41, 50, 54
    sources of ignition, 6–7
    statistics, 151–52
explosives, 30–31. *See also* permitted explosives
Falip, Jules and Henri, 103, 124, 131, 133
fatalism, 15, 21–24, 53–54
Ferguson, A.E. *See* undertakers
Fernie, William, 2,
firedamp, 5, 168n24
Frank Slide, 22, 93, 94
fraternal organizations, 96–98, 112, 137
Frolak, Yuriy, 11–13, 51
funerals
    cost of, 97–98, 101, 181n114
    marching in, 111–13
    music at, 111–12
gas. *See* methane gas
Gate, Harry, 14, 95
Genjanbre, Emile, 112
Goodwin, Albert, 12, 23
grave markers, 113–40
Grewcutt, Henry and Harry, xi–xiii, 44, 124, 126, 141–42
grief, 87, 90–93, 98, 101, 107, 132

INDEX   187

Hillcrest Coal and Coke Company, 2, 167n5
Hillcrest mine explosion (1914), xiii, xx–xxi, xxiv, 43, 44, 45, 72, 78, 82, 84, 85, 95, 113, 116, 117, 137
Homenick, Joseph, 124, 125
Hosmer, 2, *See also* mine rescue, stations
Hovan, John, 116, 120
Hutcheson, Sydney, 83, 86
Hutchinson, William, 44, 45, 57, 94
Hutton, Charlie, xii
International Coal and Coke Company, 2, 101, 167n5
Knights of Pythias. *See* fraternal organizations
Koziec, Mike, 18–19, 36
Lothian, George, 124, 128
Makin, Julia, 78, 85, 94
masons. *See* fraternal organizations
McDonald, Duncan, 45, 47
McGillivray mine, xi, 2, 51, 90, 108
memorials, 136–37
methane gas, 5–6, 10,
Michel, xiii, 40
mine rescue
  effectiveness of, 50, 55–57
  equipment, 40, 47–49, 54
  legislation and regulations, xvi, 38–41
  mining companies and, 39, 41,
  problems with, 41–46, 50
  stations, 40–41, 46
  training, 46, 50, 52–53
miners
  attitudes toward mining, 3, 9, 12, 15–29, 53–54
  code, 29–30
  experience of 9–11
  family tradition of mining, 10, 24, 26–27
  mobility of, xxiii, 13–15
mining accidents
  learned helplessness, 15
  problems when happened, 35–38
  statistics, xiii, 2–3, 151–52
Mitchell, James, 116, 121
Mitchell, John, xiv–xv, 100
Morrissey mine, 2, 167n4
Munkwitz, Henry, 10, 69
Murray, David, 44, 124
mutual aid societies, 96–98
Nycoma, Albert, 36, 37
Oddfellows. *See* fraternal organizations
permitted explosives, xv, 4–5
Picariello, Emilio, 14–15
Plasman, Henry, 123, 124
Powell, William, 28, 82
pride in work, 26–28
relief funds, 110–11
rescue. *See* mine rescue
safety lamps, xii, xv, 7–8, 168n28
Scott, William. *See* undertakers
Smith, Frank and Joe, 136–37
Smith, Robert, 124, 129
Songs, xxii–xxiii, 24–26, 29, 83, 91, 109, 136, 165n24
Special Rules, 20
Springhill mine, Nova Scotia, 22, 83, 86
St. John's Ambulance, 53
Thomson, George (Curly). *See* undertakers
Tiberg, Alrik, 12, 24
Turnbull, Matthew, 18–19
Turtle Mountain. *See* Frank Slide
undertakers
  background of, 58–59
  Davies, T.W., 59, 60, 61, 64–65, 69, 72–73
  Ferguson, A.E., 59, 61
  history of, 57–58
  Scott, William, 59, 61
  Thomson, George (Curly), xvii, 64
unions, 30–31, 61
  effectiveness of, 100, 182n120
  Gladstone Union of Fernie, 98
  hospital created by, 101–2
  Mine Workers Union of Canada, 100

One Big Union, 98
   strike action, 99, 100, 101, 104
United Mine Workers of America, xii, xv, xxiv, 30, 98–103
Western Federation of Miners, 98
wages, 99–100
Weatherby, True, 51
Western Canadian Coal Company, 2

Westray Mine, xxii, 16, 28,
Wolf safety lamp. *See* safety lamps
Woodmen of the World. *See* fraternal organizations
women
   mourning the dead, 113
   organizations of, 97
   reactions to danger and disaster, 23–25, 78, 90, 94